Time

Time is a history of the philosophy of time in western philosophy from the Greeks through to the twentieth century. Philip Turetzky focuses on the role of time in the various ontological theories throughout history, arguing that, regardless of differences in ontological theories, time always functions as a boundary condition on phenomena.

The first half of the book explores ontological theories in ancient and modern philosophy. The second half describes the philosophy of time in three twentieth-century philosophical traditions: analytic philosophy, including philosophers such as McTaggart and Mellor; phenomenology, in the work of thinkers like Husserl and Heidegger; and a distaff tradition, which Turetzky identifies as including Bergson and Deleuze.

Time is an intriguing and enlightening read. It will be invaluable to anyone who has wondered about the nature of time.

Philip Turetzky is Instructor in Philosophy at Colorado State University.

The Problems of Philosophy

Founding Editor: Ted Honderich

Editors: Tim Crane and Jonathan Wolff, *University College London*

This series addresses the central problems of philosophy. Each book gives a fresh account of a particular philosophical theme by offering two perspectives on the subject: the historical context and the author's own distinctive and original contribution. The books are written to be accessible to students of philosophy and related disciplines, while taking the debate to a new level.

Recently Published:

Time

Philip Turetzky

London and New York

First published 1998
by Routledge
11 New Fetter Lane, London EC4P 4EE

Simultaneously published in the USA and Canada
by Routledge
29 West 35th Street, New York, NY 10001

Typeset in Times by Routledge
Printed and bound in Great Britain by
TJ International Ltd, Padstow, Cornwall

British Library Cataloguing in Publication Data
A catalogue record for this book is available from the British
Library

Library of Congress Cataloging in Publication Data
A catalogue record has been requested for this title

ISBN 0-415-13948-1 (pbk)
ISBN 0-415-13947-3 (hbk)

(13914)

Contents

Contents

Contents

Illustrations

Figures

Preface

Time, as a topic of study, extends to virtually every area of intellectual inquiry and practical engagement. It impinges on the natural sciences, the social sciences, literature and the arts, politics, economics, religion, and private life. The various roles of time in these areas cannot be isolated from one another without loss. No one study, especially one of so restricted a length, can encompass this complexity. This is not surprising, since time, in any of its roles, is a fundamental aspect of all that occurs, a boundary condition on phenomena.

This book is a work in the history of philosophy. Although it is restricted to the ontology of time as it developed historically in European thought from ancient Greece through the contemporary traditions in western philosophy, the topic remains vast in scope. The difficulty in presenting this material is exacerbated by the necessity of explaining the various ontological positions in sufficient detail to make sense of the role and nature of time in each position. This has made the task of writing the book a seemingly endless series of decisions about what would have to be left out. I have not had the luxury to discuss, in detail, alternative interpretations of each position, nor why I have selected the interpretations I give. More frustrating has been the limited space for commenting on the wealth of differences and interconnections, often finely nuanced, between various positions.

In Part One, I have tried to present the various positions without pretending they are clearer or better supported than they appear. Instead, I have adopted the useful fiction that philosophical thought can be pictured as unfolding within and among the writings of philosophical thinkers. The hope is that clarifications and criticisms emerge as immanent in the history as it unfolds. Nor have I attempted to assess the strength of arguments and positions or develop an original position on the nature of time. In Part Two I am not so concerned to settle the issues regarding each tradition, as much as to place them in relation to one another through several themes. In any case, I have tried to make the great wealth of source material clear in the notes and bibliography, so that the interested reader can explore the various problems. Moreover, I have tried by judicious use of connective comments and internal references to mark divergences and connections around which multiple virtual books may haunt the actual text.

Several people have read parts of the text and helped improve it immensely, although not even such excellent help could save me from many of my follies. I wish to thank Robert Welsh Jordan for helpful comments on the phenomenology chapters, Jane Kneller for helpful comments on the Kant chapter, and William Bogard likewise on the two final chapters. I have benefited from conversations with Michael McCulloch, Garrett Thomson, Thomas Trelogan, Suzanne Unger, David Utterback, among many others. People attending lectures I have given at Colorado State University and the College at Wooster Ohio, and students in courses at Ripon College and the University of Northern Colorado have challenged me to seek greater clarity and appreciation of the difficulties the ontology of time presents. I have received encouragement and editorial assistance from Tim Crane, Emma Davis, Adrian Driscoll, Anna Gerber, Shankari Sanmuganathan and Lisa Williams. Markham Rougé helped with the drawing of some of the figures, Marie-Laure Marécaux aided in some translation from the French, and my brother, Howard Turetzky, has provided computer assistance over many years.

I must express special thanks to two people. My thanks with great affection to Adrian Moore, friend and true lover of wisdom, who first suggested that I take on such an impossible task and helped make its publication possible. Lastly, I wish to express my tremendous gratitude to Ron G. Williams, inspiration, mentor, friend, whose clarity and sense of wonder sustained me through many a dark hour. Ron read and commented on the whole of the text, making it far better than I could have hoped without his devotion of valuable time.

Part One

The history

INTRODUCTION TO PART ONE

For by time there have been and shall be brought to light all
things which were hidden.

François Rabelais

This book is an exploration of the concept of time via a proposal
about the role it has played in western philosophy. In particular, it
will examine the use of the concept of time as a boundary condition
on phenomena. This formula is not a definition of time; although in
the sense and to the extent that meaning is given by use, it gives the
meaning of time in the context of the history of ontological
thought since the emergence of western philosophy among the
ancient Greeks.

 The statement that time is a boundary condition on phenomena
intends something broad enough to be neutral regarding the great
diversity of ontological views and systems developed over the last
two and a half millennia. Yet the statement also intends something
specific enough to play a functional role in such thought. The
breadth of the statement is necessary to ensure that we avoid privileging one ontological view over another. However, this neutrality is
not without a curious ontological consequence; it makes the role of
time, in some sense, prior to considerations about the reality of time.
One task of this book is to show that the various debates about the
reality of time and about its nature presume that time plays the role
of such a boundary condition. The key terms "phenomena" and

"boundary condition" each express a family of concepts. The specificity of the statement that time is a boundary condition on phenomena requires some clarification of these expressions.

The core of the family of concepts expressed by the term "phenomena" is that phenomena are those things that appear. However, appearing can refer to the appearing of the thing itself, as when a musician appears on stage, or it can mean an appearance of something that does not itself appear, as when a disease appears through its symptoms, e.g. spots. Something can be a phenomenon in both senses: in the former case, the phenomena are the things themselves, yet in the latter case the spots themselves appear even though they may be taken as the appearance of something else. The spots are phenomena in one sense or the other depending on whether they are taken in themselves or referred to something not appearing. When an appearance is referred to something not appearing, what does not appear can be something that could itself appear, for instance a worm that causes the spots. However, what does not appear may be something that could not appear, as when an electron appears through a track in a bubble chamber, or if a metaphysical entity were manifested in a sensory object. Such complications arise partly because of different ways of distributing the honorific term "real" over what does and does not appear and partly because of variations in other ontological assumptions. So, for instance, a distinction between things manifesting themselves and a consciousness of things may be construed as expressing two different senses of "phenomena." Which sense of "phenomena" applies depends on whether and in what sense it is legitimate to consider the content of consciousness real and on what ontological claims are made about the nature of consciousness. (Such claims offer answers to questions like: is consciousness necessary for things to manifest themselves? Do things manifest themselves in consciousness directly or only in a mediated way?, etc.) For its purposes, this book must remain neutral regarding the employment of the honorific "real" and regarding the ontological assumptions surrounding what appears. By whatever means and in whatever way things appear as opposed to not appearing, we shall consider their appearances to be phenomena.

As with phenomena, so with boundary conditions. This text shall construe the concept broadly enough to include a family of alternatives regardless of where the honorific "real" is bestowed and independent of other ontological claims. At the same time, the text will retain the core meaning that the boundary conditions constitute a limit to phenomena. The nature of that limit will vary with variations in ontology, but all such variations will exhibit one of the following four boundary structures. Ontologies which posit a separation between metaphysical, or at least imperceptible, entities and phenomena may treat time as a boundary condition of phenomena:

1 as something which does not itself appear but which acts as the most immediate constraint on what does appear;
2 as itself a phenomenon which somehow encompasses and constrains all other phenomena;
3 as something neither strictly a phenomenon nor something which does not itself appear but something intermediate between the two which constrains phenomena (or mediates the relations between what appears and what does not);
4 as a double limit where two sorts of time are posited, one on each side of the boundary between phenomena and what does not appear.

Ontologies which abstain from positing the existence of metaphysical entities tend to treat time as the outer envelope of phenomena (as in 2 above). The exceptions deny the existence of time. Again, how reality is distributed over entities and which auxiliary ontological tenets are put forward are of little consequence for the limiting function of time; whether the boundary comes between phenomena and what does not appear or between or surrounding manifestations of things, time plays the role of a limiting condition even as ontologies vary. Part One will explicate the series of ontologies of time from the ancients to the nineteenth century and show how time serves in each as a boundary condition on phenomena.

CHAPTER I

Greek thought before Aristotle

For primordial Greek thinking . . . time, always as dispensing and dispensed time, takes man and all beings essentially into its ordering and in every case orders the appearance and disappearance of beings. Time discloses and conceals.

Martin Heidegger

GREEK MYTHOLOGY OF TIME[1]

The ancient Greek philosophers were primarily concerned with the problem of change: how is it possible for the same thing to become different and yet remain the same thing? Things obviously change. But change seemed to require that one and the same thing could be opposite to itself, that it could be hot and cold, wet and dry, moving and at rest; yet such opposites were incompatible. Part of the answer to this difficulty may seem obvious, that in undergoing a change something that did not change took on incompatible opposites at different times. However, at first, the concept of time was not clearly separated from conceptions of change and motion. As with most early cosmological notions, time was first personified as a mythological deity and its cosmological significance was given in origin stories. Later, time came to be treated as an aspect of the natural world. Yet even the most sophisticated of Greek thinkers always treated problems about time as subsidiary aspects of problems about change and motion.

By the time of the early mythographers Hesiod (c. seventh century B.C.E.) and Pherecydes (c. mid-sixth century B.C.E.) the Titan Kronos had become identified with time. The obvious similarity between "*chronos*," one of the Greek words we translate as "time," and "Kronos" certainly influenced this identification, as did

5

Kronos' attribute as a deity who devoured his children. But most important for our understanding of time is the myth of the mutilation of Ouranos, the sky. A basic outline of the story as told in Hesiod's *Theogony* is that great Ouranos (the Sky) hated his children and hid them in the bowels of Gaia (the Earth) until she groaned under the strain. Ouranos covered Gaia, copulating with her until their son Kronos turned on his father, castrating him with a sickle, thus separating sky from earth and allowing the other children to emerge into the gap created by Kronos' violence. Kronos threw the severed genitals over his shoulder; from the blood came the spirits of vengeance and out of the genitals sprang Aphrodite, goddess of love.[2]

If we accept the identification of Kronos with time, the image portrayed is of time opening up a gap between earth and sky in which their children emerge. Earth and sky were commonly used as images of the boundary between the human world and the divine. Homer placed Kronos at the lowermost limits of the earth.[3] And later, when, in his defense, Socrates denies that he has been concerned with divine matters, he does so by saying that he does not study things in the sky and below the earth.[4] Time, then, forms a rupture in the divine in which the world of human life and experience can appear. While this act of separation is violent, it is also generative; it produces both vengeance and love. As early Greek thinkers moved away from mythological forms of explanation to more naturalistic forms they retained in various ways the association of time with these images of a boundary on the divine, of violent rupture and change, of generation in general, and of love and vengeance in particular.

TIME AND JUSTICE: ANAXIMANDER

The first recorded mention of *chronos* (time) in Greek cosmological thought other than myth is found in a fragment of text attributed to a contemporary of Pherecydes, Anaximander of Miletus (*c.*610 B.C.E. to soon after 546 B.C.E.). Generally accepted as the earliest surviving fragment of western philosophy, it says that the principle of all things is:

> some . . . *apeiron* [unlimited] nature, from which come into being all the heavens and the worlds in them. And the source of coming-to-be for existing things is that into which destruction, too, happens according to necessity; for they pay penalty

and retribution to each other for their injustice according to the assessment of Time [*chronos*].[5]

The fragment is cryptic and much attention has been given to what was meant by the unlimited or boundless, *to apeiron*. The fragment addresses the problem of change, which, as we noted above, was the fundamental concern of ancient thought. The unlimited, for Anaximander, was the source of all existing things, which arise out of it and fall back into it; the unlimited plays the role of a substratum, which is unchanging while existing things undergo change. According to the fragment, it is necessary that existing things come to be and are destroyed. Presumably, the necessity attaches to change because existing things are delimited by their exemplifying some opposites while excluding others; hence, existing things are always limited. Whatever is limited, then, is limited by its opposite and the unlimited serves as a principle underlying these opposites; the unlimited is not subject to change because it could have no opposite. We must keep in mind here that all limited things are limited by their opposites; the only stable thing underlying these opposites is the boundless. However, this reasoning suggests that any limited thing must be subject to coming into and out of existence only if limited things are limited in time. For otherwise there could be limited things that lasted for ever and so did not change.

The last part of the fragment, "for they pay penalty and retribution to each other for their injustice according to the assessment of Time," is the part most likely to be a direct quotation from Anaximander. If we assume that it is existing and limited things which perpetrate injustice and are subject to penalty and retribution, then time somehow orders and constrains the coming to be and passing away of all existing things. Time prevents opposite things from exceeding their bounds and becoming unlimited. How Anaximander thought that this happens is a matter for conjecture, but since time makes the assessment we might assume that it ordains the duration that a given thing exists before it is destroyed or replaced by its opposite (a naturalized form of Kronos devouring his children). Empedocles, a later figure in early Greek thought (mid-fifth century B.C.E.), offered another alternative; he treated the primal opposites as imperishable and claimed that time formed a cycle from maximum mixture to maximum separation. The force of love governed the tendency towards mixture and the force of strife governed the tendency towards separation. The myth of Kronos has not yet lost its grip on these notions.

With Anaximander, that myth is most evident in the images relating to justice. In the myth Kronos achieves vengeance by separating the human world from the divine, where Anaximander thought that time orders limited things, enforcing those limits. Time dispenses justice to all things according to its assessment, which ordains that whatever may arise must also pass away. Time sets limits on the appearing and vanishing of phenomena by governing their duration, but more fundamentally time is a boundary between limited things, phenomena, and the unlimited, out of which phenomena arise and back into which they disappear.

TIME, FLUX AND THE RATIONAL PRINCIPLE OF THINGS: HERACLEITUS

The Presocratic thinkers focused on the problem of change, and time seems inextricably bound up with change. Heracleitus of Ephesus (*c.* late sixth to early second century B.C.E.) claimed that all things were in flux, yet he also insisted that this flux was subject to a unifying measure or rational principle. This principle (*logos*, the hidden harmony behind all change) bound opposites together in a unified tension, which Heracleitus likened to that of a lyre, where a stable harmonious sound emerges from the tension of the opposing forces that arise from the bow bound together by the string.

All we have of Heracleitus' writings are fragments and only one of these mentions time. It says, "time is a child moving counters in a game; the royal power is a child's."[6] The image in this fragment is that time somehow produces change within the order of nature. As a child, time is not the rational principle itself; still it is not just equivalent to changing nature. Little more can be gleaned from the text and even this interpretation may be excessive. However, it is worth noting that the word here translated as "time" is not "*chronos*" but "*aion*," from which derives the English word "eon," meaning a long duration. From the doxography we know that Heracleitus thought that time does not have a beginning; it has had an infinite duration. This seems to place him in opposition to his predecessors, who sought naturalistic cosmogonies while keeping the mythic form of an origin story. Whether Heracleitus also held the view that the world recurs in cosmic cycles is a matter of contention. (We have already mentioned that Empedocles was later to advance a cyclical view of the universe, and some form of a cyclical view would also come to be held by some Pythagoreans, Plato, Aristotle, and the Stoics; attribution of a cyclical view to

Heracleitus may have been improperly ascribed due to these later views.) If such a cyclic cosmos were compatible with the claim that time has no beginning (and if Heracleitus was consonant with his predecessors), then time had to be to some extent independent of change, i.e. it seems that it would have to be possible for changes in the whole of the cosmos to repeat at different times. Were time to be entirely dependent on or wholly an aspect of change, then nothing could count as a repetition of exactly the same cosmic cycle, for there would be no difference that would mark a difference in just the times when the cycles occurred.

Time, which to some extent is independent of change, would have been, for Heracleitus, not itself subject to changes and hence closely allied to the rational principle of things. Without a beginning, time would be of an infinite duration and so itself a feature of the world that does not change. While every thing was in flux, on Heracleitus' account, the timeless unchanging rule of the *logos* constrained and ordered the flux of things. Hence, if time were not identical with the *logos*, and it could not be reduced to a mere attribute of change either, then it somehow existed in between the rational principle and the flux of things. Much of this interpretation is speculative and the attribution of cosmic cycles to Heracleitus depends on what may be a dubious doxography. The interpretation does, however, suggest a role for time as more than just another property of changing things. The fragment credits time with power, albeit a child's power, and the lack of a beginning allies time with the unifying rational principle as itself unchanging and as encompassing the whole of things (at least in the past).

QUESTIONING THE REALITY OF TIME: PARMENIDES

Parmenides of Elea (*c.*520–430 B.C.E.) is a towering figure in the history of western thought who introduced new questions and problems into the tradition. A member of the Eleatic school, a breakaway from the Pythagorean schools of southern Italy, Parmenides is the progenitor of the discipline of logic; his critique of his predecessors invoked notions of possibility and necessity regarding the objects of discourse. The arguments Parmenides put forward so probed ontological matters that none of his successors could ignore them, for they exposed logical difficulties in the very fabric of the philosophical and scientific enterprise. Although his literary style was poetic, he presented deductive arguments and recognized that such arguments formed a coherent system; it mattered little with which arguments one started, he thought, eventually the chain of reasoning would circle back to them again. In addition to these achievements, Parmenides was the first to express and argue for a stringent form of what has continued to be a major view of time in the West, that time is unreal. Parmenides eliminated time from his ontology by appealing to the logical features of language and thought; in effect, he argued that time cannot be the object of any possible thought and consequently does not exist.

Parmenides argued for the central claim that whatever can be spoken of, thought of, or inquired into must exist, hence it is meaningless to speak or think of what does not exist; negative judgments are unintelligible. This thesis becomes the main premise in arguing against the reality of time. Parmenides writes that whatever can be spoken or thought about "is without creation or destruction; whole, unique, unmoved and perfect. Nor was it ever, nor will it be, since it now is, all together, one continuous."[7] Any object of speech or thought cannot come into being, for it would have to arise out of non-being and non-being cannot be thought or spoken of. Insofar as this is an argument against the reality of time, it argues that time is unreal because change cannot occur. The argument assumes that time is inextricably bound up with change, for it requires the premise that if change were impossible time would also be impossible. But even if we disallow this argument and accept that something can come from nothing, Parmenides argues that no reason can be given for that to occur at one time rather than another, and he concludes that objects of thought and speech must completely exist or not exist at all. Since Parmenides' basic contention rules out non-existence, whatever can be spoken or thought of must completely exist. The thrust of this claim regarding time is that none of the existence of whatever can be spoken or thought of can have ceased to be in the past or be waiting to come into being in the future. For if existence must be complete it must be "all together" and without accretions. It might be thought, at this point, that a refutation of change and an insistence on the equal reality of all existence does not refute time in the sense of duration – after all, things may endure unchanged and nothing prevents the affirmation of reality throughout that duration – but Parmenides thwarts this objection by arguing that whatever can be spoken or thought of cannot have any parts because that would introduce non-being into being. For whatever can be spoken or thought of to be divisible into parts, he claimed, would require that the parts must be different, that one part not be another part, and that requires non-being. Even unchanging duration would be divisible into parts and so cannot exist.

Parmenides' poem presented his ontology. The first part of the poem presenting the arguments we have outlined (among others) is titled "The Way of Truth." But the poem also had a second part called "The Way of Opinion." Although much of this part has been lost, the doxography suggests that it presented a cosmology along the lines of Parmenides' predecessors, constructing the world out of opposites which explain change. Clearly, Parmenides thought that all of this was false – or, worse, unintelligible – and it is decidedly odd that he would write it at all. But what is important here is that "The Way of Opinion" is a false or even unintelligible path which purports to describe and explain how things seem to ordinary people. Parmenides recognized that his conclusions rejected such views wholesale and that, given the validity of his arguments, appearances could not be real; any claim or thought about them must be false or even meaningless, for it would have to speak or think of non-being, which, according to Parmenides, is impossible. The arguments against the reality of time are allied with this abolition of appearances; in eliminating appearances and limiting them absolutely Parmenides eliminated time as well.

THE MOVING IMAGE OF ETERNITY: PLATO

Plato (*c.*428–347 B.C.E.) took seriously Parmenides distinction between appearance and reality. While he agreed that only what is unchanging can be truly real, he recognized that appearances had to be accounted for in relation to what is real. Appearances change and so, as Parmenides thought, they must partake of non-being: they cannot be completely real. Yet, Plato added, appearances also have positive characteristics, and to that extent they must also partake of reality and cannot be completely unintelligible. The Platonic theory of forms attempts (among other things) to find a way of reconciling the reality of appearances with their unreality. The theory proposes that there are hierarchically ordered degrees of reality. What Plato calls the Good is what is ultimately real and all other things derive their reality from it. The Good makes it possible to speak and think about anything to the degree that it is real; things are the subject of rational inquiry to the degree that they participate in the Good. The forms are unchanging entities that participate in the Good by being standards of perfection for entities of every kind. The forms are hierarchically ordered according to the degree to which they emulate the Good (see Figure 1). Consequently, the highest forms are those of truth, beauty, justice and other standards of goodness; lesser forms are standards of perfection for mathematical objects, while lesser still are standards for animals and physical objects. The realm of the forms is sharply separated from the realm of appearances; where the sun illuminates phenomena, making it possible to see them in a way analogous to how the Good makes the forms available to intellectual insight. The Platonic philosophy postulates the forms to explain change and multiplicity among appearances. They themselves do not change, exist always and are unique, while appearances change, come into and go out of existence, and there can be many appearances of the same thing. The forms explain changing appearances by providing something that remains the same through change. They explain the possibility of appearances of the same thing coming into and going out of existence, since they are eternal and cannot be destroyed. Finally, they explain how there can be many appearances or instances of the same thing, because they give a common standard that justifies treating different objects as instances of the same thing. For example, we are justified in treating both maples and oaks as trees because they both approximate the standard of "tree-ness" that constitutes their form.

In the dialogue Plato named after him, Parmenides deduces contradictory claims about the temporal status of the one.[8] Regarding time, he argues that if the one is in time, then it must be becoming both older than itself and younger than itself; and that it follows that the one cannot be in time for it must be neither younger nor older than itself. This curious conclusion is meant to bolster Parmenides' denial of intelligibility to appearances, for whatever is in time is an appearance. The Platonic theory of forms preserves Parmenides' sharp distinction between the realm of what can be intelligibly spoken and thought about and the realm of appearances, but does not assert, as Parmenides did, that appearances utterly lack reality. Consequently, time no longer need be entirely unreal, and Plato is able to offer an account of the nature of time that accords with his account of appearances. This account is given in the dialogue *Timaeus* as part of a cosmogony reminiscent of those given by the Presocratics and in myth. Plato has Timaeus warn us that this account is merely a likely story and that it may contain inconsistencies. This warning alerts us to accept the story only provisionally, and reminds us that the realm of appearances

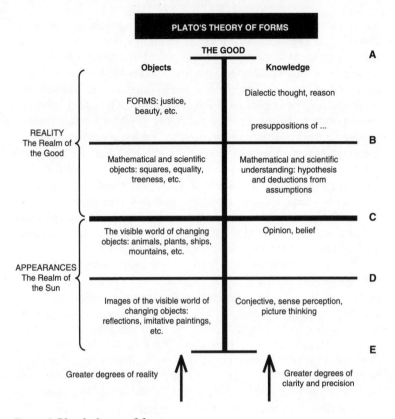

Figure 1 Plato's theory of forms

Note: The theory of forms: the separation between the realm of reality illuminated by the Good and that of appearances illuminated by the sun, and the hierarchy of degrees of reality and knowledge. The lengths of the vertical lines represent portions of reality, the longer line representing more reality. (Note that the ratios are proportional, i.e. AC/CE::AB/BC::CD/DE, and consequently that BC = CD.)

the cosmogony describes may contain contradictory (and to that degree unreal) circumstances.

The story that Timaeus tells is of a divine craftsman, the Demiurge, who creates the visible universe by imposing order and proportion on chaotic matter. Imposing proportion (*ratio* in the Greek) on matter imposes form, rationality, and intelligibility on it, thus making matter and appearances intelligible and real to the degree they embody form. The Demiurge does this by looking to the ideal forms as patterns and standards for things; as patterns the

forms serve as models, which the Demiurge copies in the multiplicity of changing appearances; as standards the forms guarantee
that the material world will be as good and as beautiful as possible.
Since the natural world is thus the best and fairest, it must be intelligent and possess a soul. For, Timaeus tells us, no unintelligent
creature could be fairer than an intelligent one. The world, then, is a
perfect animal that neither has nor needs anything outside itself to
sustain its health, since only something outside it could threaten it.
Consequently, it does not decay. We are told that its body, the
totality of matter, has a spherical shape because the sphere is the
most perfect shape; because it is most like itself – i.e. any cut
through it yields a circular plane – and it comprehends all other
figures since they may be inscribed within it. The Demiurge made
the soul and then formed the sphere within it, joining soul and body
at their centers with the soul pervading and surrounding the body.
The soul is not only the intelligence of the perfect animal but also
the source and principle of its motion, which, in keeping with the
perfection of the sphere, is circular. Because it is a fusion of form
and matter, the soul is able to judge the truth about what is the same
and what is different in the realms of both the intelligible and the
sensible. Individual souls have the same structure but less reality
than the soul of the universe. The individual soul, like the World
Soul, functions as intelligence and motive force for the individual
body, but since its body is only a part of the whole it is subject to
changes due to interaction with other bodies outside its own. This
marks a distinction between appearances as phenomena in the
world of changing things, i.e. objective appearances, and appearances as the sensory experiences of an individual being. The latter
are incomplete and imperfect in comparison to the former. For this
reason, Plato differs from modern philosophers in treating appearances primarily as phenomena in the former sense, as appearances
relative to the higher reality of the forms, rather than relative to
perceivers.

The Demiurge created the world according to the model of the
forms and sought to make the copy as much like them as possible
given the inherent limitations of the material he was shaping. To
this end he created time. Since, unlike the forms, time is created, it
belongs to the realm of appearances, a derivative order of being.
For the created universe to be as much like its eternal model as
possible, the Demiurge created time as an image of eternity. Plato
writes that time is a moving image of eternity that moves according
to number. Thus time differs from the eternity of the forms in two

respects: it moves, i.e. it partakes of change; and it moves according to number, its movement is rationally ordered. Time is an image of eternity because, while time moves and changes, its movement is the movement of the sphere of the universe. This is circular motion, in which after each complete cycle the sphere returns to its original position. We are told that "was" and "will be" apply only to processes of becoming which are in time. Time itself encompasses all such processes; it follows that time as a whole does not change and thus resembles eternity. It contains all change while it undergoes circular motion, but each complete cycle of the sphere, the great year, is as like eternity as possible, especially when the likeness between great years is considered. We are not told how closely the processes of one cycle resemble the those of the others, though clearly the motions of the heavens are thought to be exact repetitions, for their motion is itself circular and resembles eternity. While time is closely connected with the motion of the sphere, it cannot simply be identified with that motion. As an image of eternity, time links the order of the eternal world of forms and unchanging reality with the order of the natural world of changing things. This link makes the created universe into a rational, intelligible harmony that exemplifies the purpose of the Demiurge to make it as beautiful and good as possible.

Time belongs in the realm of appearances because it moves. Moreover, it moves according to number, which makes its motion intelligible. Time can be counted and regularly segmented, and so it appears as a numerical order. As a principle of order that the regular motions of bodies can measure, time is tied to the sequential order of nature. For Timaeus, it is inseparable from the regular circular motion of the body of the perfect animal. Timaeus identifies the structure of the heavens as the soul within the body of the universe. He describes the creation of the soul within the body in terms of the division of the soul substance according to definite numerical ratios. The Demiurge divided the whole mixture of the resulting proportions in two, and constructed the celestial equator (the projection of the plane of the earth's equator on to the whole of the sphere) and the ecliptic (the plane in which the planets orbit). Because the Demiurge made the former one undivided circle whose revolution determines the great year, he called it the circle of the same. The latter he called the circle of the different because it divides into the seven circles of the planets. The motions of these circles divide time, and Timaeus says they preserve the numbers of time. Years, months, days, and nights come into being only when time orders the heavens, and the motions of the planets divide time into numbers and ratios of numbers. The motions of the sun, moon, and planets measure time, and together with the circle of the same they compose an orderly universe; in this way, time orders the universe according to number.

Since time is measurable by celestial motions, it links the intelligible order of the eternal forms to the changing order of nature. The eternal forms remain static and unified, while time moves and is divisible into parts. Time is the moving image of eternity that imparts intelligibility to the celestial motions. These motions are parts of time which, by dividing it, measure it. Plato thus identifies time with the duration of the whole, and the motions of the heavens are literally parts of time. Consequently, time is co-created with the invariant motion of the sphere and of the bodies it contains. Here Plato deviates from most of his predecessors, for he claims that time had a beginning. But what does it mean for time to have had a beginning? Plato never makes clear whether there was a temporal beginning, or whether the beginning of time is atemporal and metaphysical. In the former case, there are two possible scenarios. First, prior to the creation of time, matter existed in a state of chaos with no order (harking back to early myths). While duration may have existed, it had no direction or pattern. Yet such a state would be unintelligible, since it would fall prey to Parmenides' strictures on what can be thought. Second, we might suppose that there existed a moment of time before which time did not exist, a creation of time *ex nihilo*. This suggestion also verges on the unintelligible, for what does it mean to say that time does not exist before a certain time? This "before" seems to lack sense or to assume another time in which time had its first moment. If that time were created, we would be threatened with an infinite regress. Yet, if this second time were not created, why not just treat time itself as uncreated? Alternatively, "before" might mean logically prior. On this interpretation, time had a metaphysical rather than a temporal beginning. Time is co-created not only with the orderly motion of the sphere, but with the sphere and matter itself. There would be no first moment of time. The entire temporal series would begin in the sense that it was produced by a prior cause, the Demiurge and the eternal forms. Yet Plato does not say that the Demiurge creates matter. On the model of the craftsman, the Demiurge creates by giving unformed matter intelligible form. Furthermore, if time always exists, then how is time different from the eternal forms themselves?

Plato was not clear on these matters. Later, Plotinus drew some distinctions in an attempt to clarify these issues.[9]

According to Plato, then, time is that aspect of the created world of appearances that makes it resemble the eternity of the forms. The forms articulate the Good, and the divine craftsman fashions the universe in their likeness. The circular motion of the universe integrates appearances into a single duration and makes the phenomena resemble the forms as closely as possible. The eternal forms ground time which, in turn, encompasses all becoming. It disposes appearances in the sequential order of the natural world. Time moves but does not itself become. And so time functions as a boundary condition on phenomena. The world of appearances constitutes the phenomena, and time comprises the condition of their rational and intelligible becoming. Time is the moving image of the unchanging forms that act as standards of goodness and beauty. This gives an evaluative content to time. As in the Anaximander fragment, time administers standards of justice and worth by imposing the form of the Good on phenomena.

CHAPTER II

Aristotle

what is bounded by the "now" is thought to be time – we may assume this.

Aristotle

TIME AND CHANGE

Plato was the first on record to treat time as distinct from change. Yet his account of time still bears marks of the ancient myths; it is told in narrative form as a likely story, without the support of reasoning. It is Aristotle (384–322 B.C.E.) who first provides an analysis of the nature of time and who weighs relevant arguments. Plato discussed time on the scale of the motion of the universe. He regarded time as the whole of the duration of the changing universe insofar as it conformed to the order of the forms. Aristotle, on the other hand, begins with particular changes and studies time first on the smallest scale, and only later and derivatively on the scale of the universe. His concern is to understand the nature of time, while allocating descriptions of actual duration and cosmology to other studies. To study nature is, for Aristotle, to study its principles: the principles of change and motion. So he thinks we are able to and ought to study the nature of a natural phenomenon like time over and above the actual constitution of the universe. The fundamental questions pertain to the ultimate relations between time and change, not to the actual state of duration. Aristotle begins his inquiry by asking, first, whether and in what sense time exists, and then what the nature of time is. Most of that part of Aristotle's *Physics* devoted to time examines the latter question. He says that questions about the existence of time can be resolved only by saying what time is. We will follow his treatment of the nature of time and later return to problems about the existence of time.[1]

From his predecessors Aristotle accepts the view that time is intimately connected with motion, and with Plato he holds that time has an intrinsic link with number. Examining his predecessors, Aristotle rejects their view that time is the revolution of the sphere of the universe, because the parts of the revolution would still be time even though they are not revolutions. Also, if time were the movement of the whole, then if there were more than one sphere there would be more than one time at the same time, which, he supposes, is absurd. He also rejects as naive the view that time is identical to the sphere of the universe because that view is based on the invalid argument that everything is in time, and everything is in the sphere of the whole, so time is the sphere. Lastly, he argues that time cannot be identical with motion, because each motion is always in the thing that moves, but time is everywhere. Hence, time cannot be in the moving thing in the same way motion is. Furthermore, there can be faster and slower motion, but faster and slower motions are defined by time. "Faster" and "slower" mean that the motion takes place in a shorter or longer time. So time cannot, itself, be faster or slower, nor can it be identical with motion.

However, Aristotle takes the temptation to identify time with motion as reason to think that time has something to do with motion. To settle problems such as how time can be everywhere, why there cannot be several times simultaneously, and why time cannot be faster or slower, we must examine the nature of motion. To understand time, Aristotle tells us, we need not make a distinction between motion and change. Time is involved in any sort of change, including change of place, motion in the narrow sense, alteration in quality, as, for example, when a thing changes color, generation and destruction of substances, and increase and diminution in size.

He presents a singular argument for the conclusion that time requires change. He argues to the conclusion that time depends on change from the premise that we cannot tell that time has passed if we are oblivious to change. To explain the premise he mentions "those who sleep among the heroes in Sardinia," who when they wake assume that no interval existed between the instant they fell asleep and the instant of waking.[2] This example emphasizes a point he discusses later, viz. that time may depend on changes in the soul.[3] What is peculiar about the argument is that Aristotle draws an ontological conclusion from an epistemological premise. For the argument to be valid we must generalize the case of the sleepers in Sardinia to say that time without change would be unobservable, and add that there could not be unobservable times.[4] This second claim seems to fill out the argument, but it sits ill with much of Aristotle's approach to philosophy, which does not usually endorse the move from something's being unobservable to its impossibility. We will return to the link between time and consciousness, but may suppose here that Aristotle thought that link strong enough to justify such a claim regarding time, even though he does not endorse this form of argument in general.

THE DEFINITION OF TIME

Having argued that while time is not identical with change it still must have something to do with change, Aristotle presents a definition of time by specifying the connection of time with change. Time is the number of motion with respect to before and after. As peculiar as this definition sounds at first, in it Aristotle, following his usual practice, is codifying common sense. "Motion" means change in the more general sense we have already noted, but to see the sense of Aristotle's definition we must examine what he understands by the before and after of motion and by the number of motion.

It is tempting to think that Aristotle's definition is circular because "before" and "after" seem like temporal determinations.[5] Moreover, we may be nonplused by Aristotle's explanation that "before" and "after" apply primarily to place and magnitude. Yet, we must remember that, for Aristotle, changes constitute the phenomena to be explained by an inquiry into nature.[6] We should resist our modern tendency to think of time as more fundamental than motion. (We shall see below that this tendency originates in the physics of Galileo and Newton, which resolves motion into separate spatial and temporal components.) Aristotle thinks of time as an aspect of change, and things change from something to something else. Before and after are thus formal elements of change distinct from change, rather than part of its definition. The before and after of change comprise the initial and final states of a modification of some magnitude, not just spatial magnitudes. Since change can occur between qualities, between places, as growth and diminution, and as creation and destruction, it advances from a prior state, before the change, to a posterior state, after the change. So Aristotle thinks that because "before" and "after" apply to magnitude they apply to change, and because they apply to change they apply to time.

When we perceive before and after in motion, time elapses. That time elapses indicates that it partakes of three other features of change in magnitude: extension, continuity, and transition. For there to be a before and after in change, these must be separated by some interval. In motion between places this interval is a magnitude of spatial extension, while other forms of change exhibit extended intervals proper to their magnitudes. So, for instance, a qualitative change from dark to light spans an interval of saturation. Change of magnitude and time is continuous, because magnitude in any of the dimensions of change is continuous. If it were not, then there would be no distinction of before and after in the change; the change would take no time. If a change does take time it manifests a continuous interval from its initiation to its completion. Since this interval exists between the before and after of the change, the interval is a transition from the one to the other. The change is a transition from one state of a changing thing to another. What we observe in motion is primarily the moving body, and by seeing the moving body we see its motion as a transition of the body from one place to another. Besides the moving body, its motion and the type of change it undergoes, we can also observe the time the change takes by seeing the transition with respect to its before and after. Time, then, is grounded in change, which is, in turn, grounded in the changing thing.

In this way, Aristotle's definition analyzes and refines our common-sense intuition that time covers continuous intervals of changes in magnitude. But that by itself is not enough to tell us what time must be for it to cover that interval, for the time that covers the interval is not the same thing as the interval covered. Aristotle separates these by identifying time with the number of change. This identification seems to run contrary to the usual notion that numbers are timeless. However, Aristotle points out that "number" is ambiguous between what is counted and the abstract numbers with which we count. Time as number is not abstract number, for time is the number of what is counted. One period of ten days does not have the same before and after as another period of ten days. They are not abstractly ten the way both ten horses and ten men are. Time, instead, is number in the sense of what is counted. So time is what is counted in change regarding the continuous interval of transition between before and after. This formulation more clearly articulates the ordinary conception of time.

Time is what is counted in change, but, Aristotle maintains, nothing could be counted if there were no one to do the counting.

In this way time requires the soul. Time would not exist without the soul. Movement and its before and after would exist, but they would not be numbered or counted. Note that this does not make time subjective. That a soul must exist for time to exist does not make time's existence dependent on any beliefs or mental states the soul may actually have. The connection entails only the objective requirement that something exist with the capacity to count for anything to be countable. Time is what is counted, and that requires that counting be possible. This necessitates that at least one thing exist with the capacity to count. Aristotle does not say why a being with the capacity to count must be a soul, but clearly something like memory would be necessary for prior numbers to be retained and the count to accumulate. We shall see that some later thinkers will argue for subjective accounts of time through an appeal to memory.

Insofar as time is what is counted, it can become a measure of motion. Aristotle says both that time measures motion and that motion measures time. This is not as paradoxical as it may first appear. As the number of change, time discriminates more or less change. By marking off the interval between the before and after of a movement, time can be used to compare and thereby measure motion. Yet we must take care not to confuse counting with measuring, or the countable with the measurable. Time is what is countable in motion. By counting the before and after of the motion we count the interval between. This interval can serve as a unit of measure so long as it is uniform. For this reason, regular motions, such as those of celestial bodies, measure time. In this, Aristotle's view resembles Plato's. But, given this uniformity, time can serve as a measure of other motions: whether they are longer or shorter, faster or slower. A change lasts longer if it takes more standard time units than another change, and is faster if it traverses a greater magnitude in the same standard unit as another change. Time is what is counted in change, but it measures change only by comparison with a uniform standard.[7] If there were no regular changes, change would still take time and be countable but time would not measure motion. Given a uniform motion to serve as a standard, motion measures time. Since time is the number of motion, time measures motion by counting its before and after.

THE NOW

Time is what is counted in change, and what is counted is the before and after of change. Consequently, what is counted is position in the before/after sequence; time, like change, involves a succession, and as number that succession is ordinal. Although it is by counting the before and after of change that we count the interval between, what is primarily countable are instants, or what Aristotle calls "nows." Nows are counted concomitantly with motion, because before and after are different nows with respect to time, as prior and posterior are different states with respect to motion. The now is the basic phenomenon of time, for time is the succession of nows (before and after) counted in motion. Yet Aristotle argues that time is not made up of nows, nor is the now a part of time. While there is a succession of nows in the continuum of time, the now is analogous to a point on a line. As a line cannot be made up of points, time cannot be made up of nows. Neither two points nor two nows can be next to one another; for between any two points or nows there exist an indefinite number of points or nows.[8] The analogy between nows and points raises several difficulties for Aristotle, and he discusses the now partly by spelling out where the analogy holds and where it fails.

The now, Aristotle tells us, is always the same and yet always different. As what is before and after in a change, it is always the same. Given a particular change, before and after exemplify nows as substrata. A body is the same body when it changes; likewise the now remains the same now. Aristotle says, "The now follows the moving thing."[9] In this respect, Aristotle's conception of the now resembles the concept of being present. The now constitutes a substratum that can take on different attributes, i.e. be counted first as before and then as after the change. In this sense it is always different, for, as countable, the now can be counted by different ordinal numbers. In this respect, Aristotle's concept of the now resembles the concept of an instant. Thus Aristotle thinks that insofar as a now is a now it is a point at a stage of motion, but insofar as it is countable we predicate different ordinal numbers of it. He compares this with something carried. The thing carried is a real thing that remains the same, while its being carried makes it fall under different predicates locating it in different places.[10]

The now is the same and different in another way. Like a number or a point, it both joins intervals together and divides them. The now makes time continuous by connecting the past with the future, yet it may divide time since it potentially terminates time intervals. This emphasizes the differences between geometrical points and nows. A point can begin one interval and also end another, but Aristotle claims that the mind must pause to make this distinction;

it takes time to count the point twice. If exactly the same now were counted twice, time would stand still while it was being counted. In counting the same now again, one has already passed on to a different now. The distinction between ending one interval and beginning another, therefore, must be an abstraction from the actual now. In this sense, the now is always different inasmuch as it counts changes. If the now were part of time it would be like a line or interval rather than a point, but the now is as discrete and indivisible as number.

We both count nows in changes and use nows as numbers to count time. Insofar as we count time rather than motion, nows function as numbers that do the counting. Nows are countable regarding change but do the counting with respect to time. Nows, like numbers, are indivisible, while time is infinitely divisible. So there are no minimum times. What is counted is not a single point connecting two intervals, but the extremities of intervals that note the order of the before and after of motion. Intervals or parts of time are counted only indirectly. We can say that the same now is both a beginning and an end only insofar as we abstractly consider the general structure of time. Insofar as we consider time as number, the now must mark either one or the other; the before and after of motion do not coexist. Nows succeed one another like ordinal numbers. Time intervals are continuous but time as number is not. As continuous, time is long or short; as a number it is many or few. It is tempting to say that the now before and the now after a change are the limits of the time interval. Aristotle, however, cautions us that the now is a limit in one sense but not in another. The nows are only limits of the change they embrace. They mark its starting and stopping. As limits they belong to the change or, rather, they are limits of the continuous magnitude which changes. With respect to time, however, the nows are not limits but numbers. For the nows are not parts of time and limits share the mode of being of what they limit.

Time is continuous and always in transition. While time and the now depend on one another for their existence, the now itself is not in transition. The now is indivisible and Aristotle claims that like other indivisible things it comes into existence without going through a process of becoming.[11] In what sense, then, does the now make time continuous? For there to be a continuous temporal transition there must be some change, and the change must be marked by a difference between two nows. Here again the now is always different; the earlier now must be different from the later now – it must be a different ordinal number. For before and after to count time implies a difference between the two and a definite order for their appearance. What is counted in change are nows, and this requires that the now that has been counted be seen as different from the now we are at present counting. Aristotle never distinguished between the now as present and the now as an instant. He writes indifferently of time ordered statically from earlier to later and as flowing from past through the present to the future, freely mixing static and flowing terminology.[12] He recognizes the continuous transitional nature of time, but treats the now both as the same and as always different in ways that do not appeal to a distinction between static and flowing time.

THE SCOPE OF TIME

Exploring time in its smallest confines, Aristotle concludes that there is no smallest time, because time is an interval. Time, like a line, can be divided ad infinitum. The now is not a part of time since it is indivisible and punctual, but Aristotle recognizes that ordinary usage treats "now" as an interval. Secondarily, "now" means a time near the now, strictly speaking. So "now" may designate various lengths from very short to long enough to include this year or even longer periods, where its range depends on context. Having begun with the smallest aspects of time, Aristotle, then investigates its extent. This requires answers to several questions. What does it mean to say that something is in time? What sorts of things are in time and what are not? Since time depends on motion, how can it be the same everywhere? Can time come to an end?

For something to be in time it necessarily involves time when it exists. Yet "to be in time" cannot merely mean to coexist with time. For if "to be in x" meant to coexist with x, then any coexistent things would be in each other, e.g. the universe would be in a grain of sand, which is absurd. Appealing to the definition of time, Aristotle says that, since time is the number of change, to be in time must be like to be in number. Things are in number when they can be counted with numbers or are aspects of numbers; similarly things are in time if they can be counted by time or are aspects of time. Aristotle compares being in time with being in place, where something is in place if it is contained by place. Things in time, likewise, are embraced by time, bounded by a before and an after. So, for instance, things in motion are in time because they are measured by time. Things in motion can be measured by time, because time can determine a regular interval as a unit of measurement. Moreover, even though the now is not in motion it is an aspect of time. Motions are in time as they are counted by time, but nows, e.g. before and after, are in time because they constitute time by counting it.

Generally, for something to be in time it must change quantitatively. Insofar as they are quantitative, such changes can be measured by time; they can be counted with respect to before and after, and compared with a standard time interval. Every quantitative change is in time because every change happens at some rate, faster or slower than other changes. For one change to take place at a different rate than another it must cover a larger magnitude in an equivalent interval, before and after. Since before and after count time, every change must be in time.

By definition, every change actualizes a potential. As a consequence, time measures change and the essence, or form, of that change. While Aristotle accepts the Platonic idea that the essence of a thing is its form, he finds the forms separable from the material world only in thought.[13] Since the essence of change is to actualize potential, time also measures everything merely potential. Changes not only will be in time, but also in everything that has the potential for change. Time will measure both motion and rest because only things that could be in motion can be at rest. Time intervals are continuous even if the motion they span comes in fits and starts. This applies to all forms of change. Time measures locomotion, change of size, and qualitative change insofar as they are quantitative changes. Each actualizes a potential. Moreover, while they do not actualize a potential by taking on a different place or form, things that are subject to becoming and perishing are also in time. For they can be measured by time in that there is a greater time interval that will contain, i.e. embrace, them. Finally, some non-existent things are in time insofar as they were or will be in time in the past or future. Things in time can be affected by time, in the sense that to be affected is to change and time is what is countable in change.

Time might be said to be the wisest of all things because all things come to be and pass away in time. Instead, Aristotle claims that it is more correct to say that time is the most stupid of things, because time is primarily destructive in that things change, destroying their former condition, while time is only incidentally a condition of things coming to be. A thing must itself act to come to be, but it can be destroyed without acting. Time is not an agent in production, but things decay through the passage of time alone. Consequently, anything that neither changes nor has the potential for change will not be in time, for to be in time is to be measurable by time, and time measures motion and rest. For example, geometrical relations are not in time since they are neither in motion nor at rest. Things that necessarily exist, e.g. God, and things that cannot exist, e.g. four-sided triangles, are not in time.

Time is not equivalent to change, because, while all changes are in time, one and the same time is everywhere, yet each change has a definite location. Two changes are simultaneous when they are counted by the same before and after. Changes occur at different times when their before and after differ, and overlap to the extent that their intervals overlap. These comparisons apply no matter what sort of change takes place, at whatever rate. While an aspect of change, time is distinct from any particular changes. The time of several changes is the same if their number is equal and they happen simultaneously. This is why the same time is everywhere, for "the number of equal and simultaneous movements is everywhere one and the same."[14] Ten horses and ten men are both still ten. In like manner, time is not bound to the content of change any more than number belongs to the things numbered.

Aristotle differs from Plato in that he accepts the more common Greek view that there cannot be a first or a last time. Since for any given number there is always a greater number, likewise there must be a time greater than any given time. So for any now counted there will always be another now before it and a now after it.[15] Aristotle argues that since time is always a beginning (there is always a now after any given now) time will not cease. Furthermore, time will not cease as long as motion exists, for time is an aspect of motion. While differing from Plato regarding a beginning of time, Aristotle's view most resembles Plato's when compared on the largest scale.[16] Motion can measure time if it is regular, and the most regular motion is that of the celestial sphere. It moves in a uniform circular motion, and Aristotle claims that since its number is the best known, by nature, it is the best measure of time. Other circular motions are measured by this one and are differentiated only by size. Yet time does not recur even in repeated circular motions, because the cycles each have a different before and after. Time could recur only if one and the same motion recurred. The orderly motion of the sphere is the best measure of time, but, unlike Plato, Aristotle does not use it to define time.

Time embraces all change and all potential for change, and so it encompasses all natural phenomena. For everything that has a principle of change has a nature. Time is a boundary condition for change, and so for all natural phenomena, because instants form two boundaries for each change, but they form boundaries only if they demarcate the prior and posterior of the change's duration. Hence time is a condition of the two instants forming a boundary by connecting them, for time embraces changes numbering their before and after. It might be thought that since, for Aristotle, the stars do not change they would be a counter-instance. But, while the stars do not change in themselves, they do change place with the motion of the sphere. Aristotle explicitly places unchanging things beyond the outer sphere of the stars. A deeper objection would be that the whole of time is not in time, because, though every change has a before and after, time has no first or last instant. To be in time is to be embraced by such instants. But this only shows that time is not something in time. This reinforces time's unusual status. Time is a condition of things in time. Time does not merely embrace and bound things in nature; it is a condition of natural phenomena as such.

EXCURSUS: FUTURE CONTINGENCIES

In *On Interpretation* Aristotle presents a conundrum.[17] Of a contradictory pair of statements one must be true and the other false. This rule is easily applied to statements about the past and the present, but it raises difficulties about the future. If each statement about the future is either determinately true or false, then the truth values of statements about the future are already settled, everything takes place of necessity, and there are no real alternatives. Aristotle objects to this conclusion because it makes deliberation about alternative courses of action pointless, since one of the alternatives will happen of necessity. But deliberation and action cause future events, and so future events may potentially happen or not happen. If this is so there are real alternatives.

There are many difficulties with both the formulation and the solution of this puzzle. Aristotle's own solution was to deny that future-tense statements were either determinately true or determinately false. He notes that "all existence is necessary" does not follow from "if something is the case, then it is necessarily the case." That something is necessary if it occurs does not entail that it is necessary as such. Future alternatives, he thinks, must be indeterminate. One may be more likely, but neither can be actually true or actually false. The rule of contradictories applies only to what actually exists, not to what only potentially exists.

The problem has been subjected to intense scrutiny. Most of the debate, both by Aristotle's immediate successors and by contemporary thinkers, has centered on the interpretation of the modalities of necessity and possibility, and on the status of the law of excluded middle.[18] But the puzzle also raises a difficulty about time. Does all of time exist in the same way and to the same extent? Aristotle's solution entails an asymmetry between the past and present, on the one hand, and the future on the other. The future exists only potentially; it has no actual existence.[19] The past and present in some sense actually exist or, in the case of the past, have existed. We shall turn to the question of time's existence and in what sense it can be said to exist in Chapter III.

CHAPTER III

Greek thought after Aristotle
Skeptics, Epicureans, and Stoics

But what is the voice of the present? Nothing. The present is only a point, and the voice we hear is always that of the future or that of the past.

Denis Diderot

After Aristotle, intense philosophical activity was carried on by a variety of divergent philosophical movements. This period is too often treated as one of decline. Recent scholarship has begun to rediscover the wealth and extent of these schools of thought. With respect to the philosophy of time, the approaches of Plato and Aristotle were extended, subjected to criticism, and several innovative approaches arose.

SKEPTICISM AND THE EXISTENCE OF TIME

The passages on time in Aristotle's *Physics* begin with several arguments designed to raise doubts about whether time exists.[1] Aristotle said that these arguments suggest that time does not exist or, if it does, that it barely exists. The Skeptics take up these arguments, reorganize and elaborate them, and add some additional arguments. Sextus Empiricus (c. C.E. 200) organizes the various arguments into three more extensive arguments of roughly the same form.[2] Each of the three purports to show that time cannot have either or must have both of a pair of contradictory qualities. The three qualities are being limited, being divisible, and being capable of being generated and perishing. Since any existent thing must have either each of these qualities or their contradictories (being unlimited, indivisible, or incapable of being generated or perishing), time does not exist.

Arguments concerning being limited or unlimited

If time is limited, then it begins at some time and ends at some time. But if time begins at some time and ends at some time, then there must have been times before and after time exists, which is absurd. So time is not limited. Yet suppose, instead, that time is not limited; given this supposition, Sextus adopts and expands a version of an argument from Aristotle. Part of time is past, part present, and part future. If the past and future do not exist, then only the present exists. If only the present exists, then time is limited (to the present). But the previous considerations make it impossible that time is limited. So the past and the future must exist. But if the past and future do exist, then the past and future are present, which is absurd. Hence time is not unlimited. Since time is neither limited nor unlimited, and every existent thing must be one or the other, time does not exist.

Arguments concerning divisibility and indivisibility

Since time can be divided into past, present, and future, it is divisible. But divisible things can be measured against a (standard) part of themselves. If the present measures the past, then the present will coincide with the past, and be past. If the future measures the present and the past, then the future will be present, and the present and the past will be future. But since these conclusions are absurd, time is not divisible.[3] Thus time is both divisible and not divisible and, since nothing that is both can exist, time does not exist.

Furthermore, Sextus claims that none of the parts of time exist. Again, we are given a version of an argument from Aristotle to show that past and future do not exist. They do not exist, because if they did they would be present. But the past and future cannot be present, because, then, all events would be simultaneous. Sextus then argues that the present does not exist. First, if something changes, then it changes in the present. If the present is indivisible, then something can change in an indivisible time. But nothing can change in an indivisible time. Therefore the present is not indivisible. Sextus now adapts another argument from Aristotle. If the present is divisible, then its parts must be either past, present, or future. If an existent thing has parts, then the parts must exist. But the past no longer exists and the future does not exist yet. So the present cannot be divided into parts that are past and future. Moreover, Sextus adds, because of the flux of things the present changes into the past. Consequently, the present cannot be divided into parts that are each present. Since the present cannot be divided into parts that are present or parts that are past and future, the present is not divisible. As a result, since if the present exists it will be either divisible or indivisible, and since the present is neither, the present does not exist.

Arguments concerning whether time is capable or incapable of being generated or destroyed

Part of time is past and no longer exists, and part is future and does not exist yet. So time is capable of being generated and of perishing. But since the future does not exist, if time is generated out of the future, then time is generated out of something that does not exist. Since the past does not exist, if time perishes into the past, then time perishes into something that does not exist. If something is generated or perishes, then it must be generated from or perish into something existent. Therefore time is not capable of being generated or of perishing. It is contradictory for time to be both capable and not capable of being generated and perishing. So time does not exist.

A different, more complicated argument begins by assuming that time is something that comes into being. If time becomes, then it becomes in time. For if something comes to be, it does so in time. (This need not assume a process of becoming that takes time, for, as Aristotle pointed out, the now is in time because it is an aspect of time.) Next, if one thing becomes in another thing, then that other thing must exist before the first thing can come to be in it. So if time becomes in itself, it must exist insofar as something becomes in it. Yet if time becomes in itself, then it does not yet exist insofar as it becomes, i.e. comes into existence out of non-existence. It will not exist just to the extent that it is yet to become. But if time becomes in time, then either it becomes in itself or it becomes in another time. It follows that if time becomes in itself it must both exist and not exist. Since this is absurd, time cannot become in itself. This leaves the alternative that time becomes in another time. Sextus is not thinking of another time as if there could be more than one whole time line. Instead, he considers that one part of time could come to be in another part, the present in the past or future, the past in the present or future, or the future in the present or past. In each case the part of time that comes to be in another part would not be what it was, but would, instead, be the other. Take the present as an example. If the present becomes in another time, then the present becomes either in the past or in the future. If the present becomes either in the past or in the future, then the present will be either past or future. But it is absurd for the present to be past or future. Events that are present become past, but the present itself does not become past. Similar reasoning applies to the past and future; for them to become in another part of time would be for them to *be* that other part of time. Since these conclusions are absurd, time cannot become in another time.

This leads us back to the main argument, for since time cannot become in itself, nor in another time, time cannot be generated. Yet time is capable of being generated, since parts of it do not exist. Consequently, time is neither capable nor incapable of being generated. If something exists, then it is either capable or incapable of being generated. Therefore time does not exist.

Even though the Skeptics adapt these arguments from Aristotle, Aristotle does not draw exactly the same lessons from them. Besides noting that they give reasons to doubt the existence of time, Aristotle uses them to raise problems concerning whether the now is always the same or always different, whether the now is divisible, and whether time can exist as a set of nows.[4] As we have seen, in posing the problems in terms of the now he failed to distinguish between the flowing present and a static instant. Sextus states these argument entirely in terms of the past, present, and future. While this avoids the ambiguity of Aristotle's use of "the now," it is not clear that Sextus has any better grasp of the distinction.

Aristotle presents the arguments against the existence of time as a prelude to his discussion of the nature of time. Although he never explicitly returns to answer these difficulties, it is clear that he thinks that they can be answered and that the answer depends on the nature of time. As we have already noted, his view is that time will not cease to exist so long as motion exists, since time is an aspect of motion. We may speculate that part of his answer would speak to the premise that, if the past and future exist, then they would be present. Aristotle thought that "existence" has several senses. He would have distinguished between the sense in which "exist" means "be present" and other senses. Given that his definition of time treats time as the number of change, time would exist in the way that numbers exist insofar as numbers are what are countable. We must keep in mind that such speculation, at best, provides only part of what an Aristotelian treatment of the existence of time might have been. Time is an aspect of change. As such, the question of the reality of time depends on the status of changing things.

In spite of the array of arguments, the Skeptics are no more interested in proving that time does not exist than is Aristotle. They are, however, motivated by different considerations. For them, the point of the arguments for the non-existence of time is to convince us neither that time exists nor that it does not. Skepticism is a practical philosophy concerned with leading us to a state of tranquillity (*ataraxia*). The Skeptic's method is to oppose arguments to arguments, or arguments to appearances, so as to lead us to suspend judgment on the propositions involved. Regarding time, the Skeptic supposes that sense experience inclines us extremely strongly to assume that time exists. The arguments against time's existence are marshaled to weigh strongly enough to make us doubt the propositions we are strongly inclined to affirm of the appearances. They are supposed to break the hold of the account we give of the appearances. Skeptics do not, however, doubt the appearances themselves, so long as they are acknowledged to be appearances. Consequently, they acknowledge time's appearances, but they wish us to suspend any positive judgments regarding the nature and existence of time.

THE EPICUREANS: TIME-ATOMS

Atomism originated with the Presocratic thinkers Leucippus (*c.*440 B.C.E.) and Democritus (*c.*420 B.C.E.), but it was not until after, and perhaps in response to, Aristotle's criticisms of atomism that anyone explicitly postulated atomic times. Aristotle argues that were motion indivisible a body would have to lurch from place to place, since if it moved smoothly there would be a point at which it had only partly moved into the adjacent place. But neither indivisible motions nor indivisible spaces could have parts. As a result, it would be impossible for a body only partly to traverse a given distance.[5] Aristotle also appealed to the claim we discussed earlier that time could not be made up of indivisible nows, since nows cannot be next to each other. This argument, as well as the Skeptics' arguments regarding divisibility, assumes that the now is either indivisible and not part of time, or part of time and so divisible. Aristotle thought that it followed from this that either magnitude, motion, and time were all composed of indivisibles or none of them were. Accepting this stricture that positing atoms at all commits one to time-atoms, the Epicureans advocate a sophisticated atomic theory which postulates time-atoms as indivisible durations. On this conception, time-atoms can be indivisible and yet also be parts of time that have minimal durations which combine to make up longer durations. The same considerations apply to time-atoms as to atomic magnitudes, places, and movement. They argue that things must have indivisible parts with finite magnitudes. Otherwise, things would divide infinitely into point-like entities out of which existing magnitudes could not be composed.[6] Likewise, even though atomic times cannot be divided they have some finite duration, for, as Aristotle argued, duration cannot be composed of point-like nows.

The Epicurean method argues from perception and what is perceived to what cannot be perceived. Moreover, the method holds that the attributes of what cannot be perceived are analogous to attributes of what can be perceived. The approach notes that there are lower limits to the size of things we can perceive, and concludes that the same holds true of bodies too small to be perceived. There are specks too small for us to distinguish smaller parts within them. So there will also be minimal conceivable bodies below the threshold of perception. With respect to time, this means that there are minima of perceivable time, durations shorter than which no duration can be perceived. However, such minima, while indivisible in perception, can be divided in thought on analogy with the division of perceivable durations. If we proceed analogically, these divisions will also arrive at minima. The result will be time-atoms as minimal conceivable times. Given Aristotle's stricture, minimal conceivable times can be derived from atomic motions and distances. For the Epicureans, atoms travel in the void with the speed of thought. When an atom moves an atomic distance with the speed of thought it must do so in the minimum of conceivable time. Finally, if the analogy with perceivable time is strictly adhered to, there will be times shorter than the least conceivable time. This is the time of the famous swerve, *clinamen*, of atoms as they fall through the void. Lucretius (*c.*99–55 B.C.E.) tells us that we must not think of the *clinamen* as an oblique motion.[7] Since we do not observe oblique motions that are not attributable to collisions between bodies, it would violate the analogical method to attribute such motion to atoms. Yet the *clinamen* is necessary to explain why atoms should ever collide and produce compound things. The atoms, therefore, must incline a little but "not more than the least possible."[8] This inclination is motion that takes place in less than the minimum of conceivable time.

The Epicureans therefore postulate a two-tier theory of time, one tier on the level of the perceivable and one on the level of the conceivable. The tiers are strictly analogous to one another. Each tier is bounded by its minimum, and exceeded by a transition between this minimum and something less than this minimum. All that appears to perception must fall within the bounds of the minimum perceivable time. At the level of perception, atomic perceivable durations bound perceivable times and motions, but pass over into conceivable times less than the perceivable minimum. Analogously, conceivable times exceed the boundaries of perception. At the level of conception, atomic conceivable durations bound conceivable times and motions, but pass over into inconceivable times less than the conceivable minimum. Epicurus says: "we must consider the least indivisible points as boundary-marks."[9] The *clinamen* constitutes an initial cause, a direction of motion that is imparted to bodies in a time less than that ascribable by thought.[10] Such inconceivable times exceed the boundaries of thought; all that appears to thought must fall within the bounds of the minimum of conceivable time.

While the Epicureans affirm that time comes in indivisible minima, they do not treat it as having the same ontological status as bodies. Primarily, what exists is bodies that are ultimately composed of indivisible bodies, the atoms. Bodies exhibit properties such as shape, size, and weight. Such properties are not bodies, do not exist independent of bodies, and are not detachable from bodies. Properties are not parts out of which bodies are composed, but are inseparable from the bodies themselves. The whole body owes its permanent existence to all of its properties, without which the body cannot be conceived. Yet properties exist only insofar as bodies exist; they cannot exist apart from the bodies in which they adhere. Accidents share the same mode of existence as properties, but they are not permanently associated with bodies. A body may or may not be accompanied by a given accident. Accidents cannot exist apart from their adherence in bodies. They, however, have the same sort of reality as properties. Not only do they depend for their existence on bodies, but they occur in bodies just as they appear in perception. Their nature is as it appears. Time, like properties and accidents, cannot exist independently of bodies. Yet it is not, itself, either a property or accident of things. For time has a unique form of existence; it is singular and does not share its nature with anything else. Properties and accidents that apply to other things do not apply to time. One time embraces bodies, properties, and accidents even while it depends on them for its existence. To understand time we must confront it as it is given in intuition, and not refer it to concepts or mental images. When we measure time by days and nights and their parts we perceive time as it is. It is perceived only in relation to motion and rest, and so exists only in relation to motion and rest. The intuition of time is a feeling that arises from things and what happens to them. Time's ontological status differs from that of bodies, their properties, and accidents. What happens to things, events, are accidents of those things, and events happen in time. Insofar as events occur, time must exist.[11] The occurrence of events is what appears to thought or perception. In this sense, time is a condition of phenomena.

EXCURSUS: TWENTIETH-CENTURY PHYSICS

After the Epicureans, few later thinkers accepted temporal atomism. Some Islamic thinkers of the ninth and fourteenth centuries (notably the Mutakallimun and al-Ghazali) embraced atomism and developed a theory of time-atoms. Moses Maimonides (1135–1204 C.E.) discussed these Islamic thinkers and the relation between their atomism and Greek thought.[12] He even gives an estimate of the number of time-atoms in an hour, of sixty to the tenth power. In the fourteenth century theories of time-atoms were championed by Joannes Canonicus and Nicholas Bonet.[13] These views were likely influenced by Greek thought and the Epicureans in particular, although these scholars were also aware of Indian theories of time-atoms.[14] More recently, twentieth-century physicists have considered the possibility of minimal durations in response to quantum mechanics and its quantized treatment of properties like energy, angular momentum, and spin. Should time be quantized as well? An argument that it should bears a striking resemblance to Epicurean thought on the subject. Even in antiquity, Strato of Lampsacus (head of the Lyceum, 287–269 B.C.E.) had argued against Aristotle's stricture that atomism of one property entails atomism of others. Yet in twentieth-century physics the size of the electron and considerations regarding Planck's constant suggest that there are minimum lengths. Also, the theory of relativity entails that the speed of light in a vacuum is the fastest possible speed. Dividing the minimum distance by the maximum speed yields a theoretically minimum measurable time.[15] However, a time-atom of this sort does not compare exactly with either the Epicurean minimally perceivable or minimally conceivable time. Planck's constant places objective limits on measurable quantities, but such measurements apply far below the level of perception. On the other hand, we can conceive of times smaller than those measurable because time is measured locally. In relativistic physics the measure of even a minimal duration in one place does not entail the same measure elsewhere. Two overlapping time-atoms need not begin and end simultaneously. Hence we can conceive of times shorter than that of the time-atom, even though they may have no physical significance. In fact, even though we can make sense of a minimal measurable time and assign it a theoretical value, there seems to be no physical reason to suppose that time is actually quantized.

THE STOICS: CONTINUOUS TIME

Where Epicurean materialism found atomic constituents in all things, Stoicism finds continua. Zeno of Citium (c.335–262 B.C.E.), the founder of Stoicism, defined time as the extension of movement and the measure of swiftness and slowness. This definition differs from Aristotle's, to which it bears a superficial similarity, by placing more emphasis on time's continuous extension. It also treats time as uniform regarding any motions, i.e. it measures swiftness and slowness by comparing different motions that occur in the same time. Continuity lies at the core of Stoic ontology, and its applications to the nature of time were worked out by Chrysippus (c.280–207 B.C.E.). Chrysippus amplifies Zeno's definition, adding that time is the extension accompanying the movement of the cosmos. This emphasizes not only that time is not an aspect of any particular motion but also that "it is in time that everything moves and exists."[16] Yet the Stoics do not think that time has a being independent of bodies. The Stoics adhere to a primarily materialist conception of the universe. What exists, according to the Stoics, is the mixture of bodies. But this mixture does not consist of mutually exclusive bodies, as of beans and grains of wheat, nor in a fusion of bodies, as in a chemical compound. Rather, bodies continuously interpenetrate and yet they retain their separate identities, as when many distinct sounds pervade the same air. Bodies are states of the whole, and the mixture of bodies is a superposition of states rather than a combination of atoms. Pneuma, itself a mixture of air and fire in various proportions, permeates the mixture of bodies differentiating them qualitatively. Different qualities result from different tensions in the pneuma. The universe is the total mixture of bodies. It cannot be added to or diminished, created or destroyed.

The Stoics assert that all that exists are bodies, particular substances with their individual qualities. However, the Stoics also recognize that some things have a sort of being which is not nothing at all, and yet is not fully real. These incorporeals are something, they subsist, but unlike bodies they do not exist in the strict sense. For instance, the surfaces of bodies are incorporeal. Unlike Epicurean atoms, the continuous nature of bodies requires that they have no definite boundaries. Inherently dynamic, bodies continuously pulsate. Their surfaces subsist only as limits. Chrysippus was the first to understand the concept of a limit to which an infinite series converges. The interior of bodies approaches their surfaces in a convergent infinite series. Such a limit is not part of a body, but neither is it nothing at all. It is a something, but not a body; it subsists but does not exist.[17]

Time, along with the void and things meant, is something incorporeal. It is not a body and yet it is not nothing either. Time is continuous and infinitely divisible. Chrysippus emphasizes time's continuity when he states that no time is entirely present. Subsistent time converges on the present as a limit both from the past and from the future. We may speak of the present loosely, but, no matter how small a section of time we take, part will be future and part past.[18] An infinite series of nested time intervals converges on the present as a limit. This resembles the point Aristotle made when he denied that the now was part of time. Along with this, Aristotle refused to assimilate the now to a mathematical point. For him, this was because the mind must pause to distinguish the now as the end of the past from the now as the beginning of the future. Chrysippus appeals to the inherently dynamic continuity of time and to the present as a limit, rather than to the mind's temporal passage in counting, but he would agree that any distinction between the end of one interval and the beginning of another must be an abstraction from time's dynamism. As a result, there can be no fixed points in time; events in time fluctuate dynamically. Here, the Stoic emphasis on dynamic continuity departs from Aristotle. Nows cannot serve as fixed instants in relation to which time can be counted and measured; nor can they serve as points of reference in relation to which time can measure motion.

How, then, can the Stoics claim that time measures swiftness and slowness? Without fixed temporal points, standard durations cannot be determinately fixed. The standards will always be in flux and essentially approximate. Perhaps this motivated Chrysippus to define time as the extension accompanying the movement of the cosmos. Indebted to Heracleitus, Stoic cosmology postulated recurring cosmic cycles in which the mixture of bodies is transformed into fire. In this maximum fiery state, *ekpyrosis*, the tension of air and fire in the pneuma disappears, and along with it all qualities and the cosmos (order) itself. Cosmos exists only between cycles. Without sequential order, this conflagration is not part of the continuum of time. In this way, the *ekpyroses* act as fixed points that serve as limits of the dynamic temporal continuum, and the extension of a cosmic cycle becomes a well-defined measure of rates of change.[19] Incorporeal time is infinite because it is infinitely divisible into segments, because it approaches *ekpyrosis* as a limit in which there is no last time, and because the cosmos cycles without end.

(The model of the cosmos as bounded by *ekpyroses* bears striking similarities to widely accepted twentieth-century cosmologies. These cosmologies place a singular point, where the laws of physics break down, at the beginning of the universe (the big bang). If there is sufficient matter in the universe the general theory of relativity predicts a future contraction of the universe to another singularity (the big crunch). On these models, time approaches these singular points as limits, but such points are not themselves part of time.[20])

This suggestion about time as a measure is closely related to the most perplexing aspect of Chrysippus' position on the nature of time. In seeming tension with his insistence that no time is entirely present, Chrysippus also asserts that only the present is fully real. How can we reconcile these claims? One suggestion is that, while there is no present as a point in time, the whole of one cosmic cycle constitutes a single present because it contains all that is past and future.[21] This, however, cannot be the whole story. Chrysippus says: "past time and future time do not exist but subsist; only present time exists."[22] This suggests that, in addition to incorporeal infinitely divisible time, Aion, which is always past and future but never present, there is another corporeal time, Chronos, that is entirely present.[23] Aion's infinite divisibility converges on Chronos – the dynamic present of existent corporeal time – or rather Chronos – the dynamic present – continuously divides into past and future (see Figure 2). This view is reinforced by Chrysippus' assertion that "time is to be taken in two senses . . . , namely in the sense of the whole and its parts."[24] Corporeal time exists because bodies exist; it is time referred to the whole mixture of bodies. The mixture of bodies exists wholly and all at once as completely present. It does not undergo changes, but causes them.

The Stoics analyze causality as a relation between bodies as causes and what happens to bodies – events – as effects. Everything capable of acting or being acted upon is a body, but what happens to a body – an event – is an incorporeal effect. So, for example, a knife causes the flesh to be wounded. This is marked by the present perfect participle, where "having wounded" in active voice or "having been wounded" in passive voice designates the flesh as part of the present, part of the mixtures of bodies. Marking causes with the perfect aspect presents them as already complete. Here, also, the perfect aspect is progressive or continuous, so that it denotes not just any completeness, but the completeness of a duration. The participle is a non-finite verb form (a form in which no contrast of tenses is possible), preventing the expression from marking any possibility of a contrast between past and future. Causes are thus designated as present all together, without possessing any past or future. The knife and the flesh are bodies, but the effect – the event of being wounded – is incorporeal. This is marked by the infinitive, where "to wound" designates the event. This cleaves the causal relation between the mixture of bodies – as the mutual interaction of all causes – on the one hand, and the totality of events – as their incorporeal effects – on the other. The symmetry of the mutual interaction of causes distinguishes that interaction from the asymmetrical relation between causes and effects. Bodies interact in the total mixture, but events, which happen to bodies, are incorporeal and not part of that mixture. The mutual interaction of bodies exists in its entirety as present; all bodies are co-present in the total

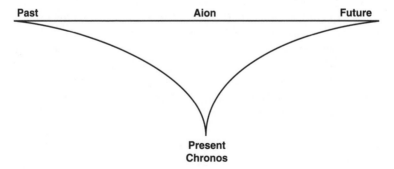

Figure 2 Chronos and Aion: Chronos is present time and not eternal because it continuously divides into the past/future time of Aion

mixture. A useful model for this situation is that of the cosmic cycle as a standing wave poised between the *ekpyroses* as nodes. The wave exhibits a stable pattern when viewed as a co-present totality, but is also an oscillating dynamic tension when viewed from a partial perspective.

Events, on the other hand, are partly past and partly future; they occur over a span of time. Just as the total mixture incorporates all bodies, incorporeal time unifies incorporeal events in the single event, the extension of the cosmic cycle. Time subsists as the event of all events. Time, then, can measure swiftness and slowness. It can do this partly because the whole span of incorporeal time provides a standard extension, but also because each event has an extension in time. The present is never part of incorporeal time, since each event is always past and future but never present. Events always occur over a span of incorporeal time. They are infinitely divisible into past and future. The full extension of this time, past and future, the event of all events, over the entire span of the cosmic cycle covers the total mixture bodies.

The Stoics are the first to entertain the possibility that there are a multiplicity of times with different characteristics. The two sorts of time are not merely two different perspectives on the same thing, the present as a whole and its division into parts. The present of the total mixture of bodies exists as the mutually interacting causes of all events. The extension of incorporeal time does not exist, but subsists as the unity of all events, past and future, as effects. In incorporeal time the present does not exist except as a limit of the infinite series of divisions. The Stoics were the first to posit two ontologically distinct sorts of time. While the Epicureans distinguished between perceivable time and conceivable time, this distinction is not ontological and the two times are perfectly analogous. For the Stoics, the differences neither mark thresholds nor are matters of perspective. The boundary between the two sorts of time occurs between cause and effect, between a present of bodies and an extended past and future of events. Time bounds the mixture of bodies as the convergence between the totality of bodies and the whole extension of the cosmos. The distinction between the two times is a condition of the causal relation, of the cleavage between the totality of causes and the unity of effects. For if the two sorts of times could not be distinguished causes and effects could not be kept separate. The corporeal present unites bodies as causes in a single cosmos. Incorporeal time unites events as effects in an infinite extension.

CHAPTER IV

Neoplatonism and the end of the ancient world

There is never more than an instant between us and nothing-ness. Now we possess it, now it perishes; and we would all perish with it if promptly and without any loss of time we did not take hold of a similar one.

                                                    Jacques-Bénigne Bossuet

The philosophical bridge between late antiquity and the Middle Ages is Neoplatonism. This return to Platonic thought and themes was conducted through commentaries on the Platonic dialogues. The interpretations in these commentaries laid stress on the cosmogony of the *Timaeus*, treating this dialogue not just as cosmology but also as an explication of ontology, an ontology that focused on the detailed elaboration of the order of beings. Discussion of time by Neoplatonists centered on interpretation and criticism of its nature and its place in that order. The best known of the early Neoplatonists is Plotinus (205–270 C.E.). We know his thoughts through *The Enneads*, six books, each of nine chapters or tractates, arranged by Plotinus' student Porphyry.[1] The discussion of time appears in the Third Ennead, Seventh Tractate, "Time and Eternity." In this section Plotinus rejects the views of his predecessors, distinguishes between temporal and timeless senses of terms like "always," explains the nature of eternity, and produces what he presents as a critically modified version of the Platonic view of time. This view of time, however, also contains emphases distinctive to Neoplatonism.

PLOTINUS AND ARISTOTLE

Like Aristotle, Plotinus critically examines the views of his prede-cessors, accepting what bears up under examination and explaining why he rejects the rest. However, Plotinus proceeds in a different order from Aristotle's.[2] He argues against the view that time is motion in general, or that it is the sphere of the universe or its motion. We have already discussed versions of such arguments. However, Plotinus adds to the thought that all motion occurs in time the notion that time is a medium distinct from motion. This presages developments that lead to Newtonian conceptions of abso-lute time.[3] The sharp distinction between time and motion gets additional emphasis in Plotinus' arguments against the view that time is some aspect of motion, its duration, quantity, or an accom-panying extension. Plotinus claims that, quantitatively, motion is measured by space not by time. All quantity and extension, he says, are spatial.[4] Here he introduces the recurring consideration that even if we were to say that time is an extension or quantity it would still remain for us to say what time is. This critique also applies to Aristotelian views that time is the number, measure or standard for measuring motion.

Plotinus' most important criticisms are directed against Aristotle. Plotinus criticizes the claim that time is a number which measures motion. First, some motions are irregular or intermittent and have no measures. Time cannot measure motion unless the motion, or some non-temporal aspect of it, is uniform. Plotinus seems to think that there could be lawless movements of this sort, and presumably that they could not be divided into lawful parts. Without unifor-mity, there could be no principle of measurement. Recall that Aristotle would have acknowledged this point, since he defined time by reference to what is counted in motion, not as its measure, on the grounds that what is counted need not be regular. Yet Plotinus thinks it is more plausible to conceive time as a measure or as a magnitude, because it is continuous, while numbers, he thinks, are discrete.[5]

Now suppose time were a magnitude providing a standard of measurement for every motion; it would have to be either a magni-tude other than and accompanying motion, an abstract number, or something else as yet undetermined. To say that time accompanies motion makes time too incidental to motion. Even though there is a tendency toward the treatment of time as independent of motion in Plotinus, it is never more than germinal. If time were a magnitude that accompanies or keeps pace with motion, it remains unclear whether the motion measures time or the time measures the motion. Precedence would depend on the nature of the magnitudes, and we have not yet determined the nature of temporal magnitude. Supposing time were a magnitude that measures motion, this does not explain what this magnitude is in itself. We would know only what time measures, but not which magnitude it is. If time were something accompanying motion, we would still have to say what this accompaniment is. It would have to come into being with, before, or after motion, so it would have to be in time.

Were time itself a magnitude, time would have to be measurable by some standard outside of that magnitude. This, again, would be either another magnitude, an abstract number, or something else as yet undetermined. Abstract numbers are discrete, while the time they are supposed to measure is continuous. Such numbers could not do any measuring. And so time cannot be measured by abstract numbers. Even if abstract numbers could measure time they would not measure time itself but only some given quantity of time or other. Here, then, we find the central consideration. Plotinus returns again and again to the point that time differs from motion, and, while it may well function as a magnitude, quantity, or number, that function is insufficient to tell us what time is in itself. He argues that time must have a nature in itself for it to provide a magnitude, quantity, or number that can serve as a measure.

According to Aristotle, time counts and measures motion regarding its before and after. For Aristotle, before and after were not primarily temporal notions. Plotinus argues that for something to be a measure of the succession of before and after it must measure according to time. He anticipates the more modern outlook by treating "before" and "after" as primarily temporal notions. Time is not brought into existence by measuring it. Things exhibit temporal succession independently of their being numbered or measured. To claim that the presence of number gives us time is like claiming that magnitude does not exist unless someone estimates its quantity. When we number or measure a section of time we presuppose time prior to its numbering or measurement.

THE NATURE OF ETERNITY

Plato defined time as the moving image of eternity. However, Plato was never entirely clear about the nature of eternity and its difference from its image, time. In preparation for his account of time,

Plotinus explores the question of what it means for something to be eternal. It is problematic what procedure to use to understand eternity. For we live in time, and we can understand matters of the eternal only by participating in them. It would seem that we ought to begin with time, in which we do participate, and only after determining its nature should we turn to the differences between being in time and being in eternity. However, Plotinus argues that this Platonic procedure of reminiscence (*anamnesis*), starting with time and moving to eternity by dialectic recollection, is not the only viable procedure. We may instead assume a common ordinary understanding of time while investigating the nature of eternity. We may, then, determine the nature of eternity and derive a deeper understanding of time by seeing how it emerges from eternity. This procedure is ontological in that it assumes the complex metaphysical structure of hypostases and emanations developed in Neoplatonism. Plotinus thinks of the eternal realm, which, for Plato, was occupied by the forms, as multiplying and expressing itself in stages. Each stage is a hypostasis, a substantial expression-form, or, rather, hypostases are substantial forms that express themselves through the emanation of additional substantial forms. In emanation, each hypostasis overflows it own bounds and produces another lesser hypostasis, weaker and containing less reality. The higher hypostasis gives up none of its reality in emanating the lower. Consequently, Plotinus proceeds by following the descent of the eternal into time through the emanation of hypostases of decreasing reality.

Plotinus recognizes three hypostases: the One, the Intellectual Principle, and the World Soul. From these the world of matter and change emanates (see Figure 3). Eternity is an aspect of all of the hypostases, since eternity belongs to real being and the structure of hypostases articulates being. Each of the three hypostases possesses eternity and each exhibits a different aspect of the nature of eternity. Eternity has being as the principle that the One possesses stability in unity. Plotinus insists that the One does not, itself, have being; it is beyond being and is the source of being. Real being implies essential existence, self-identity, and unity without change, succession, progression, or extension. Strictly speaking, only unity can be predicated of the One. No other predicates apply to the One. So, for example, while the One is also the Good, goodness cannot be predicated of it. Since unity is all that can be known about the One, only its manifestations, as approximations, are intelligible. Eternity is both unchanging and absolutely self-identical within the

One and an active manifestation of the One. Plotinus thought of the whole of reality as alive. As an active manifestation of the One, eternity is its life. Real being and real life, i.e. unchanging unity and its active principle, manifest the One.

It is difficult to understand how something can be both active or alive and also an unchanging unity. When Plato expresses the unchanging completeness of the intelligible world he says that it exists always. Plotinus chides Plato for this use of "always." When

Figure 3 The three hypostases: the One and its emanations, the Intellectual Principle and the World Soul. The changing world of individual souls and matter emanate from them.

"always" applies to unchanging reality it may mislead us into thinking that this reality has extension or that it infinitely increases in length. Eternity is without sequence, does not depend on quantity, and so has no stages. When used of what is complete and unchanging, "always" does not imply stages or extension. Instead, we should speak of being. When we speak of being always, we are noting the completeness of being in that there is nothing it lacks. Temporal things require succession; they are incomplete insofar as they take time to come to completion. In addition to this sense of "always," Plotinus distinguishes non-temporal senses of "before," "after," and "first."[6] This allows him to clear up the confusion in the *Timaeus'* difficulties concerning whether there was a temporal beginning of the material universe.[7] Plotinus avers that Plato speaks figuratively in saying that the forms existed before the universe; "before" refers only to their source or cause, not to a previous point in time.

The first emanation from the One is the Intellectual Principle or divine mind (*Nous*). This hypostasis contains the hierarchy of Platonic forms and so constitutes the intelligible essence of the world. The Intellectual Principle intuits the One and articulates its unchanging unity into the multiplicity of rational forms. While eternity hovers mysteriously about the One, it belongs essentially to the Intellectual Principle. The Intellectual Principle is an immutable comprehensive whole of all being that lacks nothing. Since eternity can be predicated both of the rational forms it contains and of the intelligible essence as a whole, the Intellectual Principle and the rational forms are eternal. Eternity is the principle of permanent existence arising from and inhering in the Intellectual Principle. That the Intellectual Principle unites the multiplicity of the forms in one being is what makes it eternal. Its eternal aspect is the unity of identity and difference, a kind of changeless activity that Plotinus calls "unvarying life." He defines eternity as the complete unchanging life of being. Eternity announces the identity of the divine in rational differences. The activity implied is the key conception of eternity as a kind of life that must be permanent, be unchanging, and lack nothing. An active being can have these features only if it exists always, in the atemporal sense Plotinus distinguishes. Eternal activity has neither future nor past; it has no extension. Since it is complete and unchanging, it is activity without past conditions or future goals. Eternal being is concentrated in an unextended unity in which inheres every possible difference. Its activity or life is manifest in the permanent synthesis and tension between its essential unity and its intelligible diversity. In eternity no difference could arise which is not already included, and so nothing new, no future, can possibly arise.

THE LIFE OF THE WORLD SOUL: TIME IN PLOTINUS

It is in the World Soul, the second emanation from the One, that time unfolds. By themselves the eternal forms have no direct connection with the material world; the World Soul mediates between the Intellectual Principle and the changing world available to sense perception. Time entails a realm of processes, the material world containing sensible things, in contrast to the atemporal realm of eternity. The sensible world is extended from the past into the future. Sensible things do not completely exist but, instead, strive for real being. In moving toward this goal their being is essentially a becoming. This movement necessitates that sensible things have a future. If the future, and hence time, were removed from the changing world, things in this world would either dissolve into non-existence or be completed and attain true being, which is atemporally eternal. Such completion is not available in the material world. The forms already possess completion. They are perfect models and standards, and hence cannot be duplicated. Since sensible things participate in existence and are not nothing they exist in time.

Time inheres in the World Soul. Time is the differentiation and movement of the life of the World Soul. The Intellectual Principle provides the immutable order of identity and difference all together, while the World Soul emanates a changing order arranging identities and differences in succession. The World Soul contains two parts. The higher World Soul looks up toward the intellectual order, while the lower World Soul acts to distribute these patterns in the changing world. This action of the lower World Soul is its life. The life of the World Soul produces changes which carry time with them; it continuously acts and so produces succession. Since, like all hypostases, the World Soul cannot contain itself; it emanates the changing world of matter and motion. This material world coming into being through successive changes is less perfect than the World Soul; it is only an image of the intellectual order. Since eternity is unchanging life, and time is life as movement and successive differentiation, the material world is the image of the intellectual order and time is the image of eternity. Time embraces the world of change and all sensible things, but it, itself, is the life of the World Soul and belongs to it, not the world of change. Time is thus coeval with the World Soul as eternity is with the intellectual order.

As in Plato, time stands at the boundary between the permanent reality of the transcendent world and the changing world of sensible things. However, Plotinus' account of time differs from Plato's in several important respects. For Plato, time operates as a vehicle of assent from the world of becoming to the world of being. Time conditions becoming by raising the chaos of becoming up to the order of the forms. Time rationally orders the world, since it moves according to number. For Plotinus, time acts as an intermediary in the descent from the perfection of the transcendent world to the mutability of sensible things. Plotinus does not accept a preexistent formless material to be molded by the Demiurge. Unformed matter has no positive existence. All positive existence emanates from the One. For Plato time belongs to the realm of appearances, while for Plotinus it belongs to the World Soul. Plato treats time as the duration of appearances. Celestial motions are parts of time because they divide the duration of appearances. In contrast, Plotinus treats time as the source of motion, as life belonging to the World Soul. Plotinus allows that, while the World Soul is not part of the changing world, its life is a kind of activity. So even though he agrees with Plato that time moves, he does not locate this movement in the realm of appearances. Time is not the duration of the movement of the sphere, but is independent of this or any other motion in the realm of appearances. Were the sphere to stop moving, the World Soul would still actively produce and order this resting state. The World Soul acts as a standard of order external to all changes of sensible things. Since he identified time with the duration of the whole, Plato thought that celestial motions were both parts and measures of time. For Plotinus, time is not a measure of motion. He argues that what serves as a measure of something cannot be its essence. So, for instance, the scale on a meter stick is not the essence of the length it measures. Motions such as the revolution of the sphere do indeed serve as measures of time, but motions do not generate time, nor are they parts of time. Motion is, indeed, necessary to make time known to us, but time is neither measure nor accident of motion. To assert that would be, he thinks, to confuse the measure with what it measures. (Plato did not actually confuse measure and measured, but his acceptance of celestial motions as both parts and measures of time encourages this error.)

Insofar as the activity of the World Soul is a kind of motion, it cannot be in time. This activity, its life, is time itself. All motion and rest are in time, but time is not in anything else. The World Soul emanates from the non-successive unity of eternity. Its life or activity, time, distributes the articulation of intelligible essence as successive appearances. Time encompasses all appearances, since the World Soul's activity produces them. Appearances would cease existing entirely if time and life withdrew. As Life of the World Soul, time is a condition of all other motion. It is the cause of the order of appearances and source of their movements. Consequently, unlike in Plato, time is not the last envelope of the changing world. Instead, while Plotinus still treats time as a boundary condition on appearances, he raises it from the upper bound of appearances to the lower bound of eternity. The world of appearances descends from the intelligible world through time, the activity of the World Soul. Time conditions appearances as the cause of their succession and source of their motion.

LATE NEOPLATONISM

Departing from Plotinus' position, the later Neoplatonists tend toward an increasingly complicated articulation of the emanation of the sensible world out of the One. Perhaps complication is inevitable given a form of explanation that generates diversity and motion by hierarchical descent from static unity. If different principles of division are assigned, any dividing point in the hierarchy can be construed as the point of emergence of a new level. Consequently, the multiplication of hypostases increases the distance between the intelligible and the sensible worlds. For, any new levels must partake of the articulation provided by the intelligible world. Iamblichus (d. *c.*325 C.E.), for example, proposed that the intelligible world is divided into three parts: the highest completely static; an intermediary level in which static elements produce dynamic elements; and the lowest level, dynamic, but still in the intelligible world. Other Neoplatonists proposed still more levels, leading to an increasingly gradual transition from static unity to diversity and motion.

Iamblichus rejects Plotinus' identification of eternity and time with life.[8] Eternity is not the static life of the intelligible world, but is itself a substantial form and, in that sense, a hypostasis. Similarly, time is not the life of the World Soul. Instead, it is a hypostasis occupying the transitional level at which the intelligible world emanates the world of change. Regarding this transition,

Iamblichus introduces an innovation of great significance in the history of thought about time. He interprets the transitional aspect of time by appeal to two different sorts of time. He distinguishes a higher time that exists as a real part of the intelligible world, from a lower time which is part of the world of change and motion, and through which participating things receive their reality. All of the higher time exists with equal reality. Its reality consists in the articulations of the intelligible world laid out in sequential order from earlier to later. Static time therefore stands at the lower bound of the intelligible world. It imparts reality to changing things that participate in the intelligible world at the point where it meets with the lower time. The lower kind of time is in constant flux, characterized by the movement of changing things from the future through the present into the past. This sort of time stands at the upper bound of the world of changing things. Changing things are unreal in the future and the past. Dynamic time tracks their motion from unreality in the future, through their coming to be real in the present, to their becoming unreal once more in the past. In this way, the two kinds of time meet at the now or, rather, there are two types of now constituting the point at which reality is transmitted from the intelligible world to the world of change. The now of the higher, intelligible, time is static, unchanging, permanent, and indivisible; it is an instant which communicates its permanence and reality to the lower now. The now of the lower, dynamic, time is flowing, changing, transitory, and continually becoming; it is the now of the real present flowing out of the unreal future and into the unreal past. The distinction between the now as an instant and the now as the present was just the distinction that Aristotle lacked in his account of the now.[9]

The material world cannot fully receive the forms from the intelligible world but only touches them at the now. The present of the passing continuum meets the intelligible world's instant and attains reality at that point. The flowing now thereby undergoes two dynamics: first, the change from unreality to reality back to unreality as participating things change from future to present to past; second, the flowing now itself moving along the static time line from earlier to later. The conception of two types of time not only distinguishes instants from the present, but also explains why "exists" and "is present" are so often identified; nevertheless, it also shows how their meanings differ. The order of time follows Aristotle's order of before and after, but things in the changing world also strive for perfection, achieving it to the degree they attain permanent presence.

Iamblichus attributes the double time view to the early Pythagorean Archytas. However, his source comes from much later writings (somewhere between 200 B.C.E. and 200 C.E.) we now attribute to a Pseudo-Archytas. These writings describe a diagram of time (see Figure 4). The Pseudo-Archytas treated this diagram as representing two aspects of a single time, indivisible with respect to stable being and unreal with respect to continuous becoming. On this interpretation, the now is a singular point of inflection which changes numerically but retains the same form. It is Iamblichus who interprets this distinction substantially as two hypostases, i.e. as two distinct types of time. Static time interprets the indivisible aspect of being, and dynamic time interprets the unreal aspect of becoming.[10] However, these interpretations must be tempered by the fact that, for Iamblichus, the higher time cannot be adequately represented by a line, or any other geometric representation, since earlier and later are parts of the intelligible world and as such have no spatial properties. He thinks of the diagram as an intellectual crutch to help us understand how the intelligible time unravels into the world of flux.

Iamblichus regards the difference between the higher time and the lower as between substantial hypostases. Plotinus wrote in terms of essential properties of time, viz. life, and accidental properties such as measuring motion.[11] Iamblichus' conception of two sorts of time departs from Plotinus' treatment of time as the life of the World Soul. Whereas Iamblichus locates the higher time and the

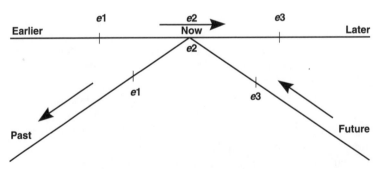

Figure 4 The diagram of the Pseudo-Archytas. The horizontal line represents point e1 as earlier than e2, which is earlier than e3. The inflected line represents e1 as past, e2 as present, and e3 as future. The now point of the horizontal line moves from earlier to later, while participating things on the inflected line move from the future through the present and into the past.

lower time on either side of the divide between the intelligible world and the changing world, Plotinus locates time in the World Soul and locates the World Soul outside the sensible world. For Plotinus, change could occur above the level available to the senses. Not so for Iamblichus; intelligible time governs the sensible world, but is itself ambivalent between rest – as it looks to the forms – and motion – as it causes the realization of participating things. Moreover, eternity was, for Plotinus, a property of the intelligible world, but for Iamblichus it is yet another hypostasis which lies at the boundary between the One and the intelligible.

The division of time into static and flowing varieties that Iamblichus offers invites comparison with the double conception of time we attributed to Chrysippus. However similar these doublings may seem, they do not split time in the same way. The Stoics treat time as a quality of bodies instead of as a substantial entity, a Neoplatonic hypostasis. The Stoic absolute separation of bodies as causes from events as effects, which prompts the separation of two kinds of time, does not resemble the Neoplatonic emanation of hypostases, which forms a chain of causes. The chain of causes judges each hypostasis to be a cause, relative to lower emanations, and an effect, relative to higher ones. Chrysippus does not distinguish earlier and later within corporeal time, but this distinction is essential for Iamblichus' higher time. The Stoics do not separate the earlier and later from the past and future. For Chrysippus all of incorporeal time has the same ontological status, but for Iamblichus the present possesses a reality lacked by the future and the past. The Stoics deny that incorporeal time contains a present, while Iamblichus asserts the existence of the now of the present and separates it from the now of the instant. The Stoics' incorporeal time divides indefinitely but does not pass, while the lower Neoplatonist time passes from the future through the present to the past. Though both schools propose two kinds of time, they accept neither the same points of division nor similar ontological specifications.

Finally, two other late Neoplatonists, Proclus (410–485 C.E.) and Damascius (c. 520 C.E.), contributed to Neoplatonic thought about time. Proclus, in his commentary on Plato's *Timaeus*, gives an argument for the existence of a changeless time in the intellectual order. He argues that motion is possible only against a background potential that is at rest. As this background potential is realized its motion requires another motionless background. Eventually, he thinks, this series must terminate in motionless time belonging to the intellectual order. Yet Proclus emphasizes the contradictory character of this higher time, which is both moving and at rest. This dual character results from the multiplication of intermediaries between the worlds of permanence and change. Like Iamblichus, Proclus argued, against Plotinus, that time could not be the life of the World Soul since inanimate things are also in time. Finally, noting Iamblichus' reservations about representing time as a line, Proclus offers an alternative image of time as a circle whose center, the intelligible world, is at rest and whose periphery, the sensible world, is in motion.

Dissenting from Iamblichus' stricture, Damascius adhered to the image of the totality of higher time existing all together as the whole of a river exists from source to mouth, except that the higher time does not flow. Time, in this respect, is like space; the entire temporal world can be surveyed, in principle, *sub specie aeternitatis*, as can the entire spatial world. Time is also like space, in that space prevents bodies from merging with one another. Space orders things which coexist. Similarly, time orders things in succession, preventing earlier events from mixing with later events. The distinction between earlier and later orders things in time once and for all, and indicates the direction of time.[12] Damascius also introduced an unusual compromise between temporal atomism and temporal continuity. He quantized physical time so that it proceeds by jumps from one unit to another, but, while each unit is isolated from the others, within each unit time is infinitely divisible. This allows the present to be concrete and real, since it exists as a unit of being and yet allows being to progress from one unit to another.[13]

CHAPTER V

Anticipations of modernity

To sum up: to stay in t_0 I must establish an objective configuration of t_0; to establish an objective configuration of t_0 I must move to t_1; to move to t_1 I must adopt some kind of subjective viewpoint so I might as well keep my own. To sum up further: to stay still in time I must move with time, to become objective I must remain subjective.

Italo Calvino

ETERNITY AND CREATION: AUGUSTINE AND PLOTINUS

The decline of the classical world and the introduction of Christian themes initiated profound changes in the philosophy of time. The Christian conscience places the highest priority on personal spiritual progress occurring within its historical narrative of creation, fall, divine incarnation, and redemption. The Christian notion of salvation depends upon this narrative of spiritual progress. Consequently, it must deny cyclical theories of time. For, were time an endless cyclical return of events the incarnation could not retain the significance of a unique saving event.[1] Moreover, classical thought conceived of time as an aspect of the order of nature, a feature of a cosmos constituted by matter and form. The Christian historical narrative and the personal salvation it promises view time primarily as an aspect of these divine and human dramas rather than as a feature of nature.

In his *Confessions* St. Augustine of Hippo (354–430 C.E.) records his spiritual autobiography. This work tells of the personal struggle which eventually resulted in his conversion to the Christian faith and his consequent rejection of fundamental aspects of his classical heritage. Augustine records his development from his hedonistic youth, through his exploration of philosophy, and finally to his conversion to Christianity. During the period he was under the sway of the classical philosophers, as a result of his study of Plotinus' writings he rejected his early Manichean belief in both a good and an evil god. He discovered, instead, the Neoplatonic conception of evil as a privation, an absence of the Good. For Augustine, Neoplatonism seemed to be the best philosophy the ancient world had to offer; only the Christian religion provided him with a more satisfactory approach to life.[2]

Plotinus and Neoplatonism had a profound and continuing influence on Augustine, even after his conversion to Christianity. Despite this influence, the Christian outlook and doctrines break not only with Neoplatonism, but also with the naturalistic assumptions of the ancient world. In the declining years of the Roman Empire, the Neoplatonic doctrine of two worlds – a real world of unchanging forms and a world of appearances in continual flux – no longer seemed adequate for personal well-being or for political stability.[3] Nor did it seem adequate metaphysically as compared with the consequences of Trinitarian doctrine given in the Nicene formula. In 325 C.E. the council of bishops convened at Nicaea by the first Christian emperor, Constantine, declared that God the Father, Son and Holy Spirit are three hypostases with one substance. By the late fourth century, under the emperor Theodosius, this formula had become entrenched in Church doctrine. As the term "hypostasis" suggests, Christian thinkers identify the persons of the trinity with the One, the Intellectual Principle, and the World Soul of Plotinus. But the Nicene formula rejects the notion, derived from Neoplatonism, that lesser reality results from the process of emanation. In making the persons of the trinity ontologically equal, the Nicene formula undermines the doctrine of emanation which had, in Neoplatonism, served to unify form and matter into a single nature. In place of a nature as the emanation or overflowing of the divine, the Nicene formula separates God, the creator, from the created natural world of form and matter.

God exists outside of the natural world, which he creates out of nothing. Augustine is aware that attributing the act of creation to God raises questions about time and eternity. He asks what God was doing before creating the natural world? Moreover, he asks, if God's act of will brought the world into being, then did his will change? Since God's will is not a created thing, but part of his substance, it could change only if his substance were not eternal. It would have to become a different substance, i.e. it would stop being what it is and so could not be eternal. But if, alternatively, God had always willed the existence of the natural world, then nature would have always existed. Augustine cannot accept this alternative. It tends to undermine the salvation narrative and it contradicts Scripture, which says there was a beginning.

In answer to this difficulty, Augustine claims that it betrays a misunderstanding of eternity. He adopts the Neoplatonic notion of eternity, which contrasts with time, even an everlasting time, were there such.[4] For Augustine, God is outside of and prior to time, just as he is outside of and prior to nature. Time itself is created, and so could not precede the original act of creation. While it makes sense to ask what happened before any given event, it makes no sense to ask what happened before all events. Since time did not exist before creation, it makes no sense to ask what God was doing during the time before the creation. In the temporal sense of "before," that would be like asking what God was doing before time. To say that God was idle before creation misconceives both time and eternity. For God is eternally present and unchanging and, as creator, he is prior to all things past and future – prior to past and future them-selves. God precedes time, but not in a temporal precedence. It is the precedence of creator to creation, and God creates nature, including time, out of nothing.

This contrasts with the Neoplatonic view, which attributes causal powers to time. For Plotinus, time was an aspect of the World Soul through which motion was communicated to the changing world given to the senses. In contrast, the doctrine of creation *ex nihilo* abjures stages of emanation. Instead, God causes all of nature to exist. Its matter and form are merely abstractions from created exis-tence. All aspects of nature, including time, are created beings dependent on God's creative power for their existence. The metaphor in the verb "depend," that created beings dangle from God, emphasizes the precariousness of created beings as effects of God's power. Both time and motion are effects depending on God's will and creative power. God's will and power are ever present, since they are required as much to sustain created beings in existence as to produce them.

TIME AND HUMAN PERSONALITY: AUGUSTINE[5]

The separation of God from nature and the doctrine that the natural universe is created *ex nihilo* wreak changes in thought so extensive that they mark the divide between antiquity and later ages. The Augustinian view rejects the classical notion of nature as merely a realm of appearances. Even though it is dependent on God's will, nature is substantial and real. Consequently, time is real because it is part of the created world. However, Augustine remains puzzled about the reality of time. His concern centers on a variation of arguments presented by Aristotle and later employed by the Skeptics to raise doubts about the existence of time.[6] Augustine notes that we speak about long and short times. However, since any interval of time can be divided into two parts, one past and one future, only the past and the future can be extended; the present cannot be extended. It must consist of an unextended instant. Were it extended, its interval would be divisible into past and future. Yet neither the past nor the future exist, since the past is gone and the future is yet to come. Only the present and what is present exist. The present is always passing into non-existence; it exists only in virtue of its pending non-existence, i.e. if the present were not passing it would not be a temporal present but would, instead, be eternal. But something can be extended only when it exists. So if the past and future do not exist they cannot be extended. The problem is that time is an extension, but it cannot be the extension of the past which no longer exists, nor of the future which does not yet exist, nor of the present since the present has no extension. It seems that time cannot be real.

Augustine accepts the force of this argument, but is unwilling to follow the Skeptics and suspend judgment when this reasoning is confronted with our experience of time. He briefly considers two possibilities. Perhaps only the present exists. This, however, fails to confront the problem, since such a present would not pass; time would not differ from eternity. Instead, perhaps, the past and future do exist but are in some way hidden from us. After all, the prophets see the future and all of us see the past in our minds. These possibilities turn Augustine's attention to our experience of time. We do, in fact, measure, count, and compare lengths of time. Instead of acceding to skepticism, Augustine works out an account that both accepts the constraints of the skeptical argument and yet explains how we are able to measure time. In doing this he accepts that only the present exists even though it passes, and that the past and future exist only in our minds.

Since we do measure time and since something can only be measured while it exists, there must be some sense in which time exists. While it exists the present (and present things) is continuously in transit towards the point where it ceases to exist. As a result, time exists only while it is passing. But the skeptical argument, applied to measurement, seems to preclude measuring time while it is passing. We measure duration between a beginning and an end. The duration of a sound, for example,[7] can be measured only once it has ceased. For if the sound continues its final point does not exist and we are unable to compare it with sounds of other durations. The past parts of the sound no longer exist and the present has no duration, and insofar as it is passing the present has no beginning or end. So it seems impossible that we could measure the duration of a sound and, more generally, impossible that we could measure time. Moreover, measuring time requires a comparison with some standard time period, but such a period can be neither past nor future, since these do not exist, and it cannot be present, since the present has no duration. One sound may last twice as long as another. To use this in measuring one by the other we must be able to compare the two durations, but both cease to exist before they can be compared.

However, Augustine sees that these arguments leave one possibility open. Since we can only measure time while it exists, i.e. while it is passing, and since duration requires some sort of extension in the past and future, there must be some way in which the past and future exist as present. If one sound follows the other, then comparing them requires that we retain the first sound while estimating the duration of the second. Past and future, then, can exist as present in our minds. The past may no longer exist, but traces of the past and of things past exist in memory images of past sense perceptions. These images are themselves present. The future and future things may not exist as yet, but we think about them, for instance when we make plans and when we predict things in the future on the basis of present perceptions, conceived of as causes or signs. So although past and future do not exist, a present of past things (memory), a present of present things (perception), and a present of future things (expectation) exist in the mind. Since these elements exist as present in the mind we can measure time by measuring their distention in the mind. The past can increase and the future decrease insofar as the mind expects, attends, and remembers. A long future, Augustine claims, is a long expectation of the future, and a long past is a long memory of the past. Memory increases in proportion to the decrease of expectation, while attention remains as what was expected passes through perception to become what is remembered. This holds true of a whole life or any of its parts.

Augustine, then, finesses the skeptical paradoxes. Although only the present is real and the present has no extension, time is real and can be measured because, as a feature of human souls, it is present and its extension is the distention of the mind in memory and expectation. Moreover, conceiving of time in this way coheres with the doctrine that the world is a creation. There can be no time before creation. Time is co-created with the world because it is a feature of created human souls. In their finitude, human beings can know the created world only by expecting things, perceiving them, and remembering them. Time, then, is a condition of knowledge.

While Augustine concedes the consequence that there will be a different time for each person, time is not subjective in the sense that it is an illusion, opinion, or governed by the will. Rather, it is a universal and essential fact about human souls that they are temporal. Thus, Augustine inaugurates a new conception of subjectivity. For the ancients, the human soul had no special status; like all other beings, it was constituted through some variation on form or matter. But this is no longer sufficient from the perspective of a created world and the salvation narrative. Humanity is created in the image of a triune God. Hence human analogues of the trinity abound in Augustine's psychology. Augustine mirrors the triune substance of God's eternal being, absolute knowledge, and divine love in the structure of the human personality, respectively as memory, intelligence, and will. For Augustine, memory is not equivalent to a storehouse of past experience, but includes a priori truths, truths of morality, self, and God; in short, memory is the storehouse of being, not merely of experience. Our contingent existence is the image of God's eternal existence. Likewise, our intelligence mirrors and depends on God's. God knows all creation without changing. Even an omniscient mind whose knowledge is like that of a human mind would have to change and attend to the passing present. God, on the other hand, is eternally present and operative in all that happens. Human knowledge is not a product of past experience, but an illumination of the human mind by the presence of the divine light. Finally, the divine will never changes, but the human will must anticipate its own issue. Instead of this future issue being governed by the will, the will itself is a phase of human temporality. Whether our expectation has issue depends entirely on God's operative presence; without divine aid we vacillate. Analogues of the trinity exist in the human soul as the temporal elements of past, present, and future, where memory constitutes the past, intelligence the present, and will the future.

That time is an aspect of the human subject emphasizes the salvation narrative central to Augustinian Christianity. It is in time that the soul is saved or lost. Since they are temporal, created beings change and so tend toward non-existence; only eternal being exists unchangeably as pure presence. Time and temporal being exist only as a present that passes, i.e. that tends not to be. The fallen state consists, in part, in a life wasted in distractions.[8] Such a life is divided between concern with what has passed and concern with what will be in the future. This span of life passing from past to future tends toward its own annihilation. So the temporal distention of the mind focuses on fleeting created beings. A soul directed toward the past or future in memory or acts of will falls into sin, because it is directed toward what lacks being. Salvation is possible only to the degree that the mind can focus on eternal things. No created thing can be co-eternal with God,[9] but by concerning itself exclusively with God the soul can achieve an ever-present state analogous to eternity. Unlike time, being ever present has neither past nor future, yet, like time, its being depends on God. Ever-presence, like eternity, has real being – a fullness. As part of the drama of salvation, created human life tends toward the fullness of time. Human beings partake of eternal life insofar as their limited nature allows, by contemplating God's eternal and immutable being, and insofar as it avoids the distractions of the world. Contemplating God perfects that limited nature and restrains human mutability.[10]

While achieving salvation through contemplation of God turns the mind from the created world, it is that world that appears in the passage of time. When Augustine denies that the natural world is merely an appearance, he emphasizes its real but dependent existence. Yet the created world appears to human beings through memory, perception, and expectation. In this way, the mind is open onto created nature, perceiving its changes in time. Time exists only in the apprehension of change by the human soul. Plotinus also located time in the soul, but, for Augustine, without the emanation of appearances from eternity time can no longer be the life of the World Soul. The doctrine of creation reconceives phenomena as the appearing of nature to the mind. Time stands at the boundary between the mind and nature as the condition of the presence of natural change to perception.

EMERGING PERSPECTIVES

Augustine concentrates primarily on the created world as a stage for the drama of salvation in history. That concern is satisfied so long as time encompasses human history and its impact on individual persons. The subjective theory of time is sufficient for the drama of salvation, but the later Middle Ages will be influenced by forces that make an account of time in the created world increasingly pressing. While Augustine does not focus on the objective foundation of time in nature, he clearly holds that time requires change. Change takes place in nature as unformed matter takes on different forms. However, change is not sufficient for time. Unformed matter is not in time at all,[11] and changes of form are only in time insofar as it is measured. While change is a necessary condition of time, the distention of the mind is its measure. Here Augustine approaches Aristotle's view, while emphasizing the role of the soul. The features that make time spiritually significant require the experiences of finite consciousness. Like Aristotle, Augustine ascribes the unity of time to the activity of the soul.

In line with Plotinus and unlike Aristotle, Augustine does not think time is an aspect of motion. Time does not belong to the physical order. There is no time in itself, only a field of relations among changeable beings, which become temporal when measured. Time cannot be the motion of the heavens alone because time measures any movement whatsoever, and such a movement may continue even should the heavens stand still or cease to exist. Moreover, because we measure movement by time, Augustine denies that time is any sort of motion. Time measures motion, and the measure of something cannot be the same as what it measures. For to measure the duration of a movement we must perceive the time at which it begins and ends. We must, then, be able to compare it with some standard temporal interval. For a temporal interval to serve as a standard it must be independent of any given motion, since time passes even though any given body remains still. While time requires some change or other, it is neither a particular motion nor an aspect of motion. After all, time measures rest as well as motion. Augustine's separation of time from motion is the first of a series of developments that produce the modern conception of absolute time.[12]

As both a divine attribute and an aspect of the possibility of salvation, understanding eternity becomes important for the medieval religious traditions. Time becomes a topic of discussion primarily through questions concerning the nature of eternity. Medieval thinkers in Islamic, Christian, and Jewish traditions drew heavily on the work of the ancients, and much of the discussion about the nature of time reiterates Neoplatonic and Aristotelian accounts. Here we should note two exceptions. The first is the temporal atomism of Abu l-Hudhayl (d. *c.*846) and others.[13] Early Islamic thought focuses on the instant of divine revelation, an indivisible moment of anguish and certainty in the human heart. This instant elapses by pointing to the future it announces; it reveals the eternal in time. This view resembles Augustine's subjective view of time. However, the condemnation of duration implicit in the revelatory instant at the heart of temporal atomism is not the dominant Islamic view. Instead, Islam generally views time and eternity as two facets of the same reality.[14] A second exception is the question of whether time is created. Thinkers like al-Ghazali (1058–1111) held that time is infinite and everlasting, and that God chose one moment for creation arbitrarily. Others, such as al-Din Rumi (b. 1207), took the Augustinian view that there is no creation in time because time itself is created as an aspect of consciousness.

Excluding the dissent of temporal atomists, most Islamic thinkers hold some variation on Aristotle's view of time. They tend to affirm the infinity of time, and to distinguish between an absolute eternal duration and the quantitative numbering of that duration. The soul numbers duration, but thinkers differ on the degree to which this makes time unreal. Al-Farabi (872–950) and Ibn Sina (Avicenna) (980–1037) hold that all divisions of duration are due to imaginative cuts of duration into instants. Later, Ibn Rushd (Averroë's) (1126–1198) treats duration as a potential in motion that becomes actual with the soul's numbering. The soul perceives time in every movement, because to perceive movement is to perceive our existence in a reality capable of change. Ibn Rushd also holds that the cause of all motion is the movement of the celestial sphere. Time is, then, an aspect of that motion. The soul only indirectly and accidentally perceives time in other movements.

The reintroduction of Aristotle's work into Christian Europe in the twelfth century from Islamic sources gave impetus to two trends culminating in the modern concept of absolute time: time begins to take precedence over motion, and time comes to be represented by an independent mathematical variable. The first of these trends resulted in the disassociation of time from motion.[15] We have already remarked on the separation effected by Augustine in making time an aspect of the soul. In contrast, Aristotle's definition of time in terms of number suggested to Albertus Magnus (1206–1280) time's independence from the soul, on the grounds that number is *sui generis*.

Consistent with the condemnation of 219 propositions issued in 1277 by Bishop Tempier, John Duns Scotus (1265–1308) posits a potential time in things that has a real existence outside of the mind (contrary to Augustine) and yet allows that it is possible for motions to continue even should the celestial sphere cease to move (contrary to Ibn Rushd's reading of Aristotle). The condemnation of 1277 proclaimed the latter an error on the grounds that God can create other worlds with other movements, and because it is possible for God to move the sphere as a whole. Potential time exists even in the absence of any movement at all. Were nature to exist without actual change, Scotus avers, it still would have the potential to act in definite ways.[16] This potential time can be known and we can measure all motion and rest by it. Time serves as a standard for the regularity and uniform speed of the succession already implicit in movement. Potential time, though objective, lacks actual existence. It has virtual existence containing distinct formal characteristics, and as the proximate cause of the actualization of these formalities in measurable time. Virtual time is, in itself, continuous and successive, but contains no actually different parts. It contains such parts only virtually; they become distinct only in their actualization. In this sense, being is univocal; it speaks with one voice. Time is a formal succession of the parts of movements, and becomes actualized as a determinate quantity when submitted to measurement. Measured time, then, is a composite of continuous and discontinuous magnitudes.[17]

Potential time is actualized in motion and quantified by measurement. The soul plays only the role of a necessary condition for measurement. Against Augustine, William of Ockham (d. 1349) argues that we can perceive time without perceiving movement in the soul, because we perceive time in any change. The turn away from the soul toward potential time, however, is not merely a return to Aristotle's view that time is an aspect of motion. We perceive time in any movement, so it only belongs contingently to the movement of the celestial sphere. This contingency consequently raises a guiding question of late medieval thought about time: does an absolute clock exist in nature? If time belongs only contingently to any given movement, then which motion should serve as the standard against which to measure all other movements? Any movement that could serve as such a standard must be continuous, regular, and invariable. The question is whether a movement of this sort exists in nature or whether our choice of an arbitrary movement as a standard makes that movement uniform by convention. Ockham concedes that the movement of the sphere may not be completely regular, but still accepts it as a natural clock. While the soul does not have to move itself, Ockham thinks that it constructs an ideal clock as soon as it perceives any change at all. This ideal clock allows us to recognize the uniformity of the diurnal movement, and so serves as a standard.

Ockham, however, offers no explanation of how a purely conceptual ideal clock can serve as a standard of measurement. Ockham's opponents argue that we choose the motion of the sphere as the standard clock by convention.[18] Since it serves as a standard, the motion of the sphere cannot be either fast or slow. Once chosen, its motion is continuous, regular, and invariable by definition; it is time and so cannot be measured by time. The choice of a standard makes it possible to determine the uniformity of other movements. Despite the conventionality of the standard clock, these thinkers still accepted the existence of natural clocks. For any movement that might be chosen possesses whatever rate of succession it exhibits due to the intrinsic properties of that movement. Not every movement can serve as a standard clock, only movements in which no physical cause can intervene and change their rate of succession. This view still ties time to motion, but it does so by separating the duration of successive motion from the standardization of measurement.

Ockham did not arrive at a pure abstract concept of time, even though he abstracted movement from particular bodies. Nicholas Bonet (d. *c.*1343) continues this process of abstraction by distinguishing between a natural and a mathematical existence of time. For Bonet, natural time exists in the duration of each movement. But while time exists in this or that movement, it only resides there contingently or accidentally. Any given movement could take more or less time, and there exists no single time in nature. There are as many times as there are movements. Even natural time is not a property of movement. The common present of two movements is extrinsic to the movements. The present of each movement consists of the coexistence of its instantaneous state with God's eternity (a view consonant with that of late Neoplatonism). Since this world is a contingent creation and other worlds are possible, no movement can claim to be first or primary. Outside of the mind each movement has its own present, and so its own coexistence with God's eternity. From the point of view of natural existence there is no unique time of all temporal things. However, from the point of view of mathematical conception there is a single time for all temporal things. The natural multiplicity of times disappears when mathematical time is abstracted from matter and motion. Time has a unified conceptual existence as a mathematical abstraction from matter and motion. The mathematical conception of time considers the successive line of time separately from all matter and movement and so conceives a single line of time, invariable and immobile. This abstraction does not falsify the reality of time, but conceives it from a mathematical perspective. The absolute clock that marks time exists, then, only as a mathematical and conceptual reality.

The second trend that culminates in a concept of absolute time is the development of the mathematical techniques that treat time as a variable independent of and prior to motion. The ancients did not analyze motion into spatial and temporal components. Even though early proponents of absolute space and time such as Strato and Boethius (*c.*480–524 C.E.) may have thought of time as independent of motion, they did not develop the mathematical tools to treat time and motion as functional variables. At the beginning of the fourteenth century Franciscus de Marchia regarded time and place as strictly analogous. He compared the relation of the time and movement of the celestial sphere to other times and movements with the relation of a container to a body placed in the container.[19] The analogy of time with space and the model of the container begin to abstract time from motion, taking a small step toward a conception of absolute time represented by a mathematical variable.

Beginning with Thomas Bradwardine (fl. 1328), thinkers at Merton College, Oxford, worked on the geometric treatment of intensive qualities. Bradwardine is the first to use the mathematics of proportions systematically to express physical relationships. This approach was further developed by other Mertonians, along with a concept of instantaneous velocity. Rejecting the priority the ancients attributed to motion, they conceived of velocity as a intensive quality and expressed it as a ratio of the independent variables space and time.[20] Nicole of Oresme (d. 1382), who is best known for developing the theory of impetus, worked out a geometrical treatment of intensive qualities, or what he called the "latitude of forms." Changes in a quality can be expressed as quantitative additions of the same quality and represented by a line. The rate of change in a velocity, for example, can be represented geometrically by representing the extension of time by a base line, a longitude, and erecting perpendicular lines of different lengths, latitudes, on it to represent different intensities of velocity. The curve connecting the summits of these perpendiculars describes the changes in intensity.[21] Time, then, is an enduring succession whose extension can be represented by a line. It is continuous and does not vary in intensity. Velocity becomes a dependent variable which can vary in intensity with respect to time. This does not completely free time from motion. Oresme still holds that time is an aspect of motion, viz. its continuous succession. But, this geometrical treatment of time as a variable independent of velocity anticipates Cartesian coordinates. When combined with Bonet's notion of abstract mathematical time, it begins to take on the characteristics of absolute time in Newtonian mechanics.[22]

EXCURSUS: SOCIAL AND HISTORICAL FACTORS

Several sociocultural factors in the Middle Ages abetted the gradual emergence of the new notion of absolute time. The division of the day into hours appears in Benedictine monasteries, beginning as early as the sixth century. The requirement that monks collectively devote themselves to prayer as much as possible makes for more prayer periods than are marked by daily natural events. *The Rule of St. Benedict* specifies a schedule dividing the daylight into eight canonical hours. All other activities then begin to be regulated by these hours. First in approximately 40,000 monasteries and gradually in other arenas, schedules begin to regulate life. Schedules divide the day into abstract units, making abstract time a matter of habitual practice. The schedule gradually becomes a major tool in the regulation of life.[23] However, the change in the length of daylight made the length of the hours vary according to the change of season. To compensate, it became necessary to vary the schedule accordingly. This necessity is closely linked to the development of mechanical clocks, since changes in light and temperature have little effect on them, and their independent regularity abolishes the need for variations in the schedule.

The first clocks operated by verge and foliot escapement mechanisms came into use in the monasteries and cathedrals of Europe in the fourteenth century.[24] They are the first devices to mark equal abstract units of time with precision. It is unlikely to be a coincidence that conceptions of abstract uniform time develop in the same historical period as mechanical clocks. Mechanical clocks measure time independently from the movement of the celestial sphere. This encourages the view that any motion can serve as the standard measure of time. Clearly, it is significant that the debate about a natural standard of time is discussed in terms of the existence of an absolute clock. Moreover, mechanical clocks facilitate precise determinations both of what time it is and of how much time has passed in a given period. This encourages the tendency to use spatial metaphors referring to lengths of time etc. The capacity of mechanical clocks to facilitate quantitative comparisons between different temporal intervals also invites this tendency to represent time geometrically.

The new ability to mark time with greater precision becomes an issue in labor disputes in the textile industry of the fourteenth century. Employers began to pay workers according to work time rather than by the piece, and conflicts arise over the imposition and accuracy of the timekeeping.[25] But the economic contentiousness of time has earlier roots in Augustine's conception of time as a divine creation in the soul. The Benedictine rule accords with Augustine's notion that salvation requires the maximization of the time spent in contemplation of God's eternity through prayer. That time belongs to God; that it is sinful to waste it in distractions points to the sinful nature of usury. Usury became a major issue in the thirteenth and fourteenth centuries. The Church sees it as a crime against God, because to collect interest on money loaned just because time has passed is a form of theft. By collecting interest the usurer steals time which belongs to God. He makes a profit without contributing anything productive, instead selling time which does not belong to him.[26] Hence economic issues of labor and profit invest the subjective aspects of time, while schedules and clocks develop the objective and abstract mathematical aspects.

TRANSITION TO MODERNITY

Due to the emphasis on continuous quantities, temporal atomism never catches hold in the late Middle Ages. The exceptions to this are followers of Ockham, Nicholas Bonet and Joannes Canonicus (c. 1329). Their atomism allows them to consider instantaneous states of motion. Most other thinkers, from Bradwardine to Galileo, could conceive of instantaneous velocities only counterfactually, in terms of what would happen should a movement continue in one way rather than another. Even Bonet and Canonicus acknowledge the continuity of time, and what is significant is that they each posit a strict separation of natural from conceptual time. Bonet holds that real time in nature is discontinuous, has indivisible parts, and a foundation in movement, while in the mind time is conceived as continuous and infinitely divisible. Canonicus reverses this assessment, holding that time in nature is continuous and flowing, and that the mind renders this flux into discontinuous cuts.[27] In any case, the duality of time posited by both Bonet and Canonicus reflects the growing separation between the natural world of bodies in motion and subjective states.

With respect to the philosophy of time, the Middle Ages mark a transition from the ancient to the modern world. We have seen many of the elements of modern approaches emerge in Neoplatonism, notably the beginning of treatment of time as separate and prior to motion. As with any transitional period, portions look back to previous positions, while elements of the later stage begin to emerge. These changes coalesce in characteristic themes of modern philosophy. For instance, the interest of modern thought in subjectivity emerges in the subjective time of the soul described by Augustine. Augustine claims that time is independent of motion, that subjective time provides an absolute standard for measuring motion, that time consists in a distention of the soul, and that the physical world changes and so moves from one state to another in succession. The last claim is difficult to reconcile with the earlier claims, and it is only with the development of the modern conception of objective time as both an independent variable in motion and an absolute flow that the two can be made consistent. The objective mathematical time of Newtonian physics grows out of the gradually loosening ties between time and motion, and the development of mathematical techniques for describing nature.

CHAPTER VI

Absolute and ideal time

Depth of space, allegory of the depth of time.

Charles Baudelaire

For René Descartes (1596–1650), the separation between objective bodies and subjective states becomes a dualism of different kinds of substances. Each of these substances can manifest only modifications of an essential property: bodies essentially have extension and minds essentially think. Although Descartes proposes a distinction between time and duration, these are not distinguished simply according to the split between mind and body. Substances, both mental and physical, have duration. Duration is the degree to which a created substance persists in its being. God's causal activity is necessary, not only for a created substance to come into being, but also for it to endure.[1] On the other hand, Descartes takes the traditional view of time as the measure of motion, but sees this as adding "nothing to the notion of duration ... but a mode of thinking."[2] Time, therefore, is subjective, an abstract mode of thought that measures the duration of created substances which may cease to exist at any moment. This distinction between time and duration prefigures problems about the ideality and reality of time that pervade the debates of the modern era.

ABSOLUTE TIME: NEWTON

The distinction between mathematical and natural time in the work of Nicholas Bonet, and his and consequent treatment of time as a mathematical variable independent of motion set the stage for the development of the concept of absolute time in the work of Isaac

Newton (1642–1727). Descartes developed the coordinate geometry of space, but he did not fully appreciate the geometrical aspects of time. Yet Galileo (1564–1642) had already represented time geometrically by means of a line marked off in regular intervals.[3] This geometric representation of time, like the uniform regularity of the mechanical clock, encourages the analogy of time with space and reinforces the priority of time over motion. For Newton, time is neither the number of motion nor the duration of things, their amount of being. Time has its own nature, independent of the existence or non-existence of the created world.

Newton's tutor Isaac Barrow (1630–1677) articulates a view of time that anticipates much of the Newtonian account.[4] Barrow says that time implies neither motion, nor rest, nor consciousness; it flows regularly regardless of the existence or state of the natural world. While we can only perceive time via some measure, perception is not necessary for time to exist. The measurement of time requires motion, and since time flows uniformly only simple uniform motion can serve as a measure of time. Any such motion will do, though we first use motions near to us, only turning to celestial motions later because of their regularity. In accord with Galileo, Barrow treats time as a homogeneous geometric magnitude. Time, he claims, is either the path of a continually flowing indivisible instant or a single succession of instants spread out in length. In either case, it may be represented by a mathematical straight line. Time existed before creation of the world; it extends from the infinite past into the infinite future. Because this is so, time is not an actual existence, but a potentiality of permanent existence. This potential time is, for Barrow, the eternal duration of God.

Newton, however, does not adopt Barrow's view that time is the possibility of permanent existence. He says that time is neither substance nor accident, but has, instead, its own manner of existence. Time is an "emanent effect of God."[5] Time is real, but it has the peculiar status that it is a consequence of the existence of God; it is impossible for God to exist without time existing. Newton holds that anything that exists must exist at some time or other. In contrast with Plotinus, even the divine existence is temporal, and since God necessarily exists he must exist at some time, namely always. God has this temporal location, but as an emanation time is also an effect of God's sempiternal existence.[6] Newton argues that time is uncreated, that it is caused by emanation from God's existence. As an emanent effect of God's existence, time arises from the fact of God's existence, and God has the property of temporal location since he exists at every time. God endures for ever, while his essence and existence are causally, although not temporally, prior to the existence of time. As in Neoplatonism, time emanates from God's existence, and so is not identical with God. God creates the world of nature and, contrary to Leibniz's accusation,[7] he is not a world soul, but the creator and ruler of nature. Time, in turn, precedes nature and is co-temporal with God. All created beings exist and move in the infinite time of God's existence, but have no effect on God; nor are they affected by his omnipresence. Things could not differ at different times save in the infinite time of God's existence. Time, then, acts as a boundary condition between God's person and the created world of nature. It is identical neither with God nor with nature, and yet is a necessary condition for the existence of nature and natural processes.

Newton distinguishes between absolute, true, mathematical time and relative, apparent, common time, separating time from its sensible measures. He defines time as follows:

> Absolute, true, mathematical time, of itself, and from its own nature, flows equably without relation to anything external, and by another name is called duration: relative, apparent, and common time, is some sensible and external (whether accurate or unequable) measure of duration by the means of motion, which is commonly used instead of true time; such as an hour, a day, a month, a year.[8]

This definition alters the ontological status of Bonet's natural and mathematical time.[9] For Newton, mathematical time is real time and not merely a conceptual construct or an abstraction from matter and motion. Newton conceives of true time as flowing in a single mathematical line, independently of all matter and motion. While we need to rely on objects that appear to the senses as measures, absolute time exists regardless of the state or existence of such objects. Time is neither a cause of nor affected by events or states of nature.

Absolute time flows equably. Without any dependence on natural objects, mathematical time flows at a constant and continuous rate. The rate of change of all natural processes and quantities must be measured against the uniform flow of time.[10] Yet in what sense can mathematical time be said to flow at all? The notion that time flows indicates, at least, that time is an irreversible succession.[11] Moreover, until the nineteenth century the calculus was conceived as tracing kinematic changes in variables.[12] Given the importance for Newton's physics of continuity and its representation in the differential calculus, we may speculate that Newton follows Barrow's suggestion that time is the path of a single indivisible instant. Barrow's other conception of time as a succession of instants spread out in length would, then, describe the temporal path trailing the flowing instant analogous to a path in space. In the Newtonian system, each indivisible moment of duration exists everywhere, while the whole of space exists at each moment. Since all motion occurs against the uniform rate of time, the whole of space and its contents develops from instant to instant or, rather, in the path of the flowing instant.

Since relative time depends on the relation between time and sensible objects, Newton asserts that, although relative time is useful in practical matters, it should not be used in the study of nature. Relative time depends on the uniformity of actual motions, even though no truly uniform motion may exist. The rate of motion may increase or decrease, but the flow of absolute time remains constant. Measured by reference to observable movements, relative time may therefore be deceptive. Consequently, it is only an appearance. Still, Newton claims that we deduce the existence of absolute time from its sensible measures. The sensible measures of time can be derived from regular motions, such as those of the pendulum clock or the eclipses of the moons of Jupiter. Unfortunately, the sensible quantities of movements may differ from their real quantities. It seems as if true time would have to be measured with respect to an absolute motion which is not accessible to observation.

All we can observe are the differences between movements, i.e. relative motions. Absolute motion must be generated by a force, while relative motion does not require one, for a force may have been applied to the other bodies with respect to which a given body is in relative motion. Moreover, a force always changes the true motion of a body, but need not change its relative motion since a similar force may also be applied to the other bodies. If bodies change their motions with respect to one another, then some of them must undergo some absolute motion, but we cannot determine by direct observation which ones or in what quantity. However, Newton notes that in real circular motion bodies tend to recede from the axis of motion, while in relative circular motion no such effects can be observed. If a bucket of water is suspended and set in circular motion the surface will become concave as real causal forces act on the water. However, if the bucket is in motion only relative to the circular motion of some other object, say a rotating platform on which an observer stands, then the surface of the water will not change. This indirect observation of real forces by means of their effects empirically establishes the difference between absolute and relative motion.[13]

Similarly, absolute time can be distinguished from relative time by observation of the effects of absolute motion. A body undergoing absolute motion must take some amount of absolute time. We can, then, derive absolute time intervals from an analysis of these absolute movements. Further, the existence of absolute time can be deduced from the correction of less regular movements, such as the diurnal motion of the sun, by more regular motions. By coordinating and comparing these motions, we may determine which are the more regular. Better, more complete and extensive, observations will produce improvements in time measures and increasingly tend to approximate absolute time. In this way, absolute time serves as an ideal to which time intervals derived from measurements of regular motions approximate, and to which they converge. This provides empirical grounds for accepting absolute time intervals.[14]

Here we have another way in which Newtonian time is a boundary condition on phenomena. Mathematical time is a condition and ideal limit of relative, apparent – i.e. phenomenal – time. As they improve, time measures approach this ideal limit empirically. It is also approached theoretically through the unity of time. All things exist in the order of succession of time, and this order of succession is immutable, i.e. absolute motion and rest must be referred to absolute times that always retain their positions in the order of succession with respect to one another. To the extent that motion and rest take place within the time interval taken by longer processes, they occur within the same time. Each process takes place within the same time interval as those processes that contain it, forming a series of time intervals of increasing length, culminating in a single time containing all the others. Hence, each temporal interval is part of one and the same time. This will be the absolute time interval for the whole of the natural world. If the natural world forms a continuous whole of all motion and rest, and all phenomena appear in this world, then absolute time will contain the natural world and constitute a condition of phenomena appearing in it.

The inheritors of Newtonian physics soon separated absolute time from theological considerations. As concern with theological matters waned, the importance of time for the definition of the new physical concepts increased. Physicists described absolute time as the temporal container of motion and rest, and the ideal limit of the order of succession. Regardless of the container metaphor, time functions in physics as one of the fundamental factors in the system of natural laws and causal forces. The new conception of natural laws, and the causal forces they govern, links empirically available phenomena by precise mathematical relations between variables. In this system of laws and forces time operates as a fundamental quantity and motion becomes a function of time as an independent variable. Time as an independent mathematical variable explicates and analyzes the new conception of motion, and so defines physical concepts such as velocity, momentum, acceleration, and force as well.

LOCKE AND EMPIRICISM

While the measurement of time depends on observation of regular motions, the concept of absolute time these measures approximate as an ideal limit must itself be empirically grounded. In *An Essay Concerning Human Understanding*, John Locke (1632–1704) attempted to provide an epistemological grounding for the concept of time.[15] While Newton distinguished true mathematical time from its sensible measures, Locke's problem is to show how from its sensible measures we can construct an idea of true mathematical time. Although it is congruent with Newton's natural philosophy, Locke's epistemology concerns how our idea of time represents real time rather than the relation between relative and absolute time. Yet if real time is Newton's absolute time Locke must show how we can form that idea and not just an idea of apparent time.

Time is rooted, for Locke, in our idea of duration, i.e. time is measured duration. Following the basic tenet of his epistemological project, Locke's method is to trace the idea of duration to its original sources in sensation and reflection. Duration, Locke claims, is a sort of length derived from transient parts of the succession of ideas in the mind. Ideas – both sensations and the complex ideas constructed from sensations and reflection upon those sensations – come into and disappear from the mind. The idea of succession derives from reflection on this passing train of ideas in the mind. The idea of duration derives from reflection on the distance between ideas in this succession. The idea of duration must derive from reflection, since we have no sensation of duration itself. Moreover, since Locke denies that a single idea has any duration, the idea of duration requires that a sequence of ideas be given in reflection and bound together in memory.[16] Should the succession of ideas cease, as in dreamless sleep, so will our perception of duration, regardless of events in the external world. However, once the idea of duration is acquired from internal reflection it can be applied to external things. Our reception of a succession of ideas constitutes our duration, and that duration is the same as the duration of things coexistent with us. This allows us to understand that external events continue and to estimate their duration, even though we slept through them.[17]

The idea of succession does not arise from observing motion, for a perception of motion occurs only when something produces a continuous sequence of distinguishable ideas, and thereby an idea of succession. There is no idea of motion in the mere fact that an idea appears in the mind. However, experience shows that it is not possible to hold a single idea long in the mind without any variation. We perceive motion only insofar as it causes a succession of ideas, and the ideas of succession and duration can arise without any motion at all. We do not perceive extremely slow movements, because their changes cause new ideas in us only after a long interval. Likewise, extremely fast movements do not produce distinct ideas of their motion. They do not cause a sequence of ideas in us, but only a single idea. Locke says that we call that part of a duration in which one idea passes and in which no succession is perceived an "instant." Since movements that are very fast or slow still have successive parts, we may not perceive their succession. Our perceptions of them are limited. Locke thinks it likely that these limits are similar in different minds. We understand very slow movements or those swifter than an instant only by inference.

Time is the measure of different lengths of duration. It seems impossible to measure duration since no part of duration is permanently available as a standard, the way the extension of a ruler can be laid alongside other extensions. Any universally observable regular appearances may serve as a standard. What matters in a measure of duration is periodicity; without regular periodic repetitions time could not be measured. The measures of duration can never be demonstrated to have exactly equal successive lengths even though duration itself is constant and uniform. The best we can do is make probable estimates of the equality of successive durations. The succession of ideas is the standard for all other successions. So the regularity of these periods must be estimated by the passage of ideas in our minds.

Time should not be confused with motion. Motion is useful as a measure of duration only because it can be uniformly periodic. Since equality of period may occur with any recurring appearance, any periodically recurring idea would do as a measure of time. Moreover, Locke thinks it strange to claim that time is the measure of motion, since he thinks of motion as translation of position which depends on space and mass as well as duration. Even were motions so irregular that they could not measure duration, the idea of duration would remain. We may imagine duration extending even were no motion to exist, by imagining the repetition of a given time interval. We may hold a time interval in memory and apply it to an empty duration, just as we can, in thought, apply a given length to measure extensions where there is no body. The duration we measure need not be coexistent with what we use as a measure. After all, the idea of duration itself requires that we hold in memory a sequence of ideas that do not coexist. All we need is the clear idea of a standard length of time and the capacity to repeat it. Given a length of time held in memory as a standard, we can continue to add this idea to itself. We derive the idea of eternity from the fact that we can imagine repetition of duration indefinitely. The general idea of time derives from consideration of any part of the idea of eternity. Whether the world had a beginning in time or existed eternally is irrelevant. Since we can conceive of either, we may suppose either alternative.

Locke's epistemology of time is the mirror image of time in Newton's natural philosophy. Locke clarifies duration as what Descartes called a subjective mode of thought. The idea of duration corresponds to and represents the objective duration of Newton's absolute time. While subjective duration represents absolute time, for Locke relative or apparent time mediates between the two as their measure. We have seen that, for Newton, we deduce the existence of absolute time from its sensible measures, i.e. from relative time. Apparent time approximates absolute time as observable motions can be shown to be increasingly uniform. This role appears in mirror image when Locke argues that observable uniform motions provide approximate standards for the measurement of duration, but the standard for all succession is the succession of ideas in the mind from which the idea of duration, truly constant and uniform, derives. Apparent time conditions all measurement of duration. Apparent time, then, acts as a boundary condition between the subjective idea of duration and the objective duration of absolute time. The approximation of absolute duration stands at the boundary of natural phenomena, on one side, and at the boundary between the phenomenal world and the perceiving subject, on the other. In this way, this conception resembles the two-tiered view of the Epicureans.[18]

Later empiricists add little to Locke's account of the ideas of duration and time. George Berkeley (1685–1753), however, objects to Locke's notion that the idea of duration represents a non-appearing objective duration. Berkeley tries to take Locke's epistemology more seriously than Locke, and argues that the Newtonian notion of absolute time cannot be derived from the succession of ideas. Berkeley holds that time is identical to the succession of ideas.[19] Any attempt to form an abstract idea of a uniform public time must, he thinks, raise insurmountable difficulties. Such a doctrine would necessitate that there exist intervals in which the mind does not think. Berkeley finds this prospect absurd, because he thinks that the mind must think to exist at all. Furthermore, Berkeley thinks that it is possible for the rate of succession of ideas in one mind to differ radically from the rate of succession in another mind. Hence, a different time will exist for each mind; however, he never makes clear what sense can be made of different rates of succession of ideas if time just *is* that succession. Finally, Berkeley agrees with Locke that one idea directly follows another, with no idea intervening. So if time is the succession of ideas, he argues, it cannot be infinitely divisible, as objective duration is supposed to be.

RELATIONAL TIME: LEIBNIZ

Gottfried Wilhelm Leibniz (1646–1716) argued against both Newton's conception of absolute time and Locke's derivation of the idea of duration from the succession of ideas.[20] A real existent absolute time would violate two principles: the principle of sufficient reason, which Leibniz holds to be a fundamental principle of reason on a par with the law of non-contradiction; and the identity of indiscernibles, which Leibniz holds to follow from the principle of sufficient reason and to apply to all substances. The principle of sufficient reason states that nothing exists, no event occurs, and no truth holds without there being a reason adequate to determine why it is so and not otherwise. The identity of indiscernibles is the principle that states that no two substances can be exactly alike, differing only numerically. If two such substances were to exist, God would have to act without a sufficient reason for creating them both.[21]

Applying this sort of argument to time, Leibniz argues that there could not possibly be an absolute time of the sort Newton advocates. Were time absolute, things could exist at any time. There could be no reason for things to exist, or for events to occur at one time rather than another. Appealing to the principle of sufficient reason, Leibniz asserts that there could be no reason for a thing to exist or for an event to occur at a particular absolute time. Newton disclaimed any causal efficacy for time. As a consequence, the absolute time an event occurs cannot enter into the determination of the event. While there must be a cause for every event, there can be no cause for it occurring at a given absolute time; there could only be a reason for an event to succeed its cause. Moreover, if time were absolute the created universe as a whole, with its order of succession remaining constant, could have been created at a different absolute time than it actually was. Leibniz objects that there can be no reason for creation to occur at one absolute time rather than another.[22] According to the principle of sufficient reason, God must have some reason to act one way rather than another. Such reasons must be grounded on some difference between the possible actions. Since empty absolute times are exactly alike there cannot possibly be any reason for God to create the universe at any particular absolute time.

This suggests, in addition, that the exact likeness between absolute times also violates the identity of indiscernibles. Were the universe created at some absolute time, time would have to have been empty for an infinite duration before the creation of the universe. Empty time would consist of infinitely many identical times. So were absolute time to exist, there would have to be infinitely many indiscernible times, which, according to the identity of indiscernibles, would be identical. The time intervals and instants of empty absolute time could not be numerically distinct as the Newtonian theory requires.

While time may be empty in the sense that there can be time intervals without any changes, an empty time is ideal, an abstraction from the relations among changing things. A recent argument shows that it is possible to give a relational account even to the claim that the whole universe may undergo definite periods without change.[23] Suppose the universe were divided into three parts such that part one undergoes a period without change every three years. The temporal interval of such a change is determined relative to changes in the other two parts of the universe. Suppose also that part two undergoes a similar period without change every four years and part three does likewise every six years. There will be a period without change in parts one and three every six years, as determined in relation to changes in part two. However, supposing the intervals begin together, at twelve-year intervals there would be a determinable time in which no change occurred in the universe. This unchanging period can be abstracted from the relations between the cycles in each part of the universe. This changeless period will, however, take a definite quantity of duration relative to changes in the universe as a whole. Hence the proposal could not apply to an infinite time before creation. Like any relation, an unchanging time interval can have a definite quantity because it is abstracted from a more inclusive order. Nor need relations be real things to have quantity, since ordinal series can have quantities.

Leibniz also offers two less compelling considerations against absolute time. He claims that God is the only infinite substance. If time were an infinite substance it would have to be God. But time has parts and God does not, so God and time cannot be identical and time cannot be an infinite substance. Moreover, it is an error to say that infinite time is God's eternity, since that would mean it is part of God's essence. To say, with Newton, that time is a property of God, Leibniz thinks, is an abuse of words. For if God exists in time that would amount to saying that "a subject is in its property."[24] Second, Leibniz holds that all true propositions are analytically true. Were this so, all false propositions would be contradictions. But it is no contradiction to say that time is relative to the order of created things.

Against Locke's account of the origin of the idea of duration, Leibniz argues that no train of perceptions is constant enough to correspond to the uniform passage of time. A succession of perceptions may arouse the idea of duration in us, but does not create it. Leibniz accepts Aristotle's notion that time is the number of motion. We know a duration by the number of regular periodic motions that occur in its span. The duration of a given period of motion can, then, be used as a standard of measure. He thinks Locke is right to say that observed motion can only be approximate. We cannot observe successive events, but only observe bodies in more or less uniform motion. However, we can deduce uniform periods of motion from the laws of nature and rules that govern non-uniform motions. (Here Leibniz is closer to Newton's successors than is Newton.) The span of duration is ontologically prior to its limits, but we can notice it only through limited extents striking our senses. To Locke's claim that we may hold a time interval in memory and apply it to empty durations, Leibniz replies that this just shows that time applies to possible as well as actual existents. In this respect, time and duration are like eternal truths. The concept of eternity results from indefinitely continued *possible* progressions derived from an abstract rule of addition of like intervals.

Like Aristotle, Leibniz holds that instants are not parts of time but the termini of temporal intervals. Temporal intervals exist only as the relational properties of actual existents. Leibniz's rejection of absolute time implies that time is not an entity distinct from whatever exists in time. The essence of matter consists in forces of resistance; duration arises because forces act and continuously succeed one another. Forces dispose material bodies in an order of succession that is logically prior to and constitutes temporal relations. Time, then, consists in the system of relations inhering in this order. Antecedent circumstances in the order of forces contain the ground of consequent circumstances. This order of succession conditions the temporal order. In this way, all existing circumstances can be ordered as coexistent, or before, or after one another in time. Multiple existing concrete circumstances that are not mutually exclusive are contemporaneous, while non-contemporaneous circumstances may include incompatibilities. Time, as a whole, is the order of non-contemporaneous circumstances; the abstract universal order of change. It includes all durations. The duration of a thing is its quantity of time, how long it lasts, minimally an instant of zero magnitude. Each thing has its own duration, but not its own time, for things change in time, but the duration of a thing remains through its changes. Duration remains the same through a succession of incompatible states of things where those things are altered by some cause.

Temporal intervals, then, are relations: abstractions from the relational properties of things. Leibniz considers them to be *ens rationis*, beings of reason, abstracted from the real relations among things. Unlike Locke, who believes there are real relations external to whatever stands in those relations, Leibniz holds that relations are internal; they have no reality when considered in abstraction from the things that stand in those relations. As such, time intervals can be divided indefinitely. Indivisible parts of time cannot exist; all that exists are instants which are not parts of time. Instants consist in the boundaries of these intervals, abstract cuts from the order of things. All that is real are substances and their properties, including their relational properties. Since relations are abstractions from real things, abstractions from the relational properties of real things, they are unreal.[25] The relational theory of time stands opposed to Newtonian absolute time since the relational theory implies that time is ideal and not a real thing.

THE ACTIVITY OF SUBSTANCE: SPINOZA AND LEIBNIZ

For Leibniz, reality consists of substances, their attributes, and modifications of those attributes. His views diverged from his great predecessor Baruch Spinoza (1632–1677) in that Spinoza argued that there could be only one substance, which Spinoza calls "God or Nature," while Leibniz sanctioned an infinite number of substances or monads. God is only one of these monads. All other monads are creations of God existing only contingently, while God exists necessarily. For Spinoza, on the other hand, substance as such must be infinite and can exist only necessarily. Every finite or limited existent is merely a modification of the attributes of God or Nature. In spite of these differences regarding substance, there is an important similarity between the two positions that has a bearing on the nature of time. Leibniz and Spinoza both think that substance is a unified multiplicity out of which articulated multiplicity arises. Substance has no parts and yet gives rise to infinite diversity that expresses its essence. Only phenomena can be divided into parts. The essence of substance is already a multiplicity that can be expressed and articulated into parts only through the substance's attributes and their modes.[26] In this sense, substance actively produces the phenomenal world. What appears is, Spinoza contends, the expression of infinite substance through infinite attributes. For Leibniz, both infinite substance – or God – and finite substances, the other monads, actively express their nature in the perception of phenomena.

Spinoza conceives of duration, along lines similar to Descartes, as the indefinite persistence of existing things. As such, duration can be distinguished from the existence of a thing only by a distinction of reason. Time is the measure of duration derived from comparison with some standard motion. Duration is indefinite, because the duration of a thing is not determined by the nature or essence of that thing.[27] Because the divine existence follows from the divine essence, God or Nature has no duration but is atemporally eternal. Whatever has duration persists through change; God or Nature does not change but, retaining its simple unity, expresses itself in infinite diversity. In this diversity causal series concatenate antecedent grounds with their consequent effects. Yet such series must themselves be grounded in God or Nature. Leibniz would say that there has to be a sufficient reason for God to select one possible chain rather than another. For Spinoza, however, God's causality is immanent in the causation of each effect in the chain of causes. Spinoza emphasizes the distinction between superficial causal chains and substantial duration when he discusses quantity in general. He distinguishes between quantity as given to the imagination – finite, divisible, and made of parts – and quantity as substance known by the intellect – infinite, indivisible, and one.[28]

In this sense, the productive activity of substance is of a different sort than the causal relations of forces acting in succession. This activity resembles Neoplatonic emanation, except that the series it produces express the essence of substance and that its effects do not possess a lesser degree of reality than the activity that produces them. For Newton, the emanation of time more closely resembles emanations in Neoplatonism, since it undergoes a diminution of being. For Spinoza, time as produced by the activity of substance is an expression and articulation of the very being of that substance. Spinoza says that God or Nature is a "substance consisting of infinite attributes, each of which expresses eternal and infinite essence."[29] The essence of a substance contains everything that ever happens to it; its nature is fixed, whole, and exists in a unity or co-presence. This co-presence or eternity expresses itself in the articulation of its real internal multiplicity. The activity of substance does not presuppose that there is a time in which the activity occurs.[30] Substance is active because it is a simple unity containing a real multiplicity that cannot be divided into parts.[31] The phenomenal world really and actually expresses and articulates this fixed unified whole. Duration and its measure, time, articulate the existence of phenomena according to series of grounds and consequents. Duration and time, then, condition the phenomenal world, which can only express the eternity of the active internal principle of substance in articulate attributes. Phenomena must be laid out in articulate series in a successive order. Without this limitation the inherent multiplicity of substance could not be expressed. The manifestation of substance requires time as a boundary condition so that its successive order may express incompatible modifications. The implicit multiplicity of substance can, then, appear as multiple but articulated phenomena.

CHAPTER VII

Kant

Over several centuries, from the Greeks to Kant, a revolution took place in philosophy: the subordination of time to movement was reversed, time ceases to be the measurement of normal movement, it increasingly appears for itself and creates paradoxical movements . . . time is no longer subordinate to movement, but rather movement to time.

Gilles Deleuze

A major tendency in the philosophy of time culminates when, in his *Critique of Pure Reason*, Immanuel Kant (1724–1804) completes the inversion in priority of movement over time. This inversion deeply influences western thought afterward, and not just thought about time. Time takes on an importance for philosophy in the nineteenth and twentieth centuries that it did not have before Kant. Until Kant's inversion other aspects of philosophy involved the philosophy of time, but afterward positions on the nature of time tend to shape whole philosophical orientations. While Newton privileged space over time, he prepared the ground for Kant's overturning of this order of precedence by making time independent of change. Kant finds fault with the accounts of time given by both Newton and Leibniz. He argues that these accounts fail because they assume that time is a feature of things as they are in themselves, independently of our capacities to grasp them.[1] He accepts Leibniz's criticism that the Newtonian account illicitly treated time as an infinite self-subsistent container of events, because, on that account, time can stand in no conceivable relation to the things and events it contains. The Newtonian conception of a real absolute mathematical time conflicts with the principle that nothing actually infinite can be given empirically. For Kant, the advantage of the Newtonian view was that it allowed for the application of strict mathematical laws to things and events given in experience. In contrast, Kant thinks that since Leibniz treated time as ideal relations of succession abstracted from phenomena he must deny that mathematics could apply validly to real things in time.

THE NATURE OF TIME

Kant offers a middle way between these alternatives. He argues that time is ideal, but that the concept of time is not derived from sense experience alone. We experience real objects outside us, but our knowledge of these objects is not merely inferred from our sensations. Like the empiricists, Kant insists that all possible knowledge of objects must be tied to and constrained by sense experience. However, he rejects their view that all such knowledge must be derived from sense experience. He argues that certain conditions must hold for any experience of objects even to be possible. These conditions must be provided a priori as universal and necessary presuppositions of any possible experience of objects. The conditions of any possible experience of objects transcend the contents of sense experience, and yet because they are necessary for experience of objects to be possible they can be known to us only insofar as we can demonstrate that necessity. Kant calls such conditions "transcendental," and since he thinks that they constitute ideal capacities or concepts conditioning the experience of objects by finite rational beings he entitles his position "transcendental idealism." Every experience, and all knowledge of objects, presupposes transcendentally ideal conditions that make it possible to know empirically real objects. Whatever things may be like in themselves, for us to experience them or have knowledge of them empirical objects must conform to the structure of our capacities to experience and know them.[2] Transcendental idealism, then, claims that conditions of the possible experience of objects are also conditions for the objects themselves to be given as appearances.

Time has the structure of the absolute mathematical flow of Newtonian time, and likewise cannot be discovered through any sense experience; time is an a priori condition of any possible experience. Kant's fundamental characterization of the nature of time is that time is "the formal *a priori* condition of all appearances whatsoever."[3] Any way in which an object can be directly given for cognition Kant calls intuition.[4] Intuitions present particulars, about which we can make judgments by bringing them under concepts. Concepts are universal rules that unite different appearances of the same object. The only way that human beings can stand in unmediated relation to empirical objects, i.e. have intuitions, is through sense experience. Taken solely as given in experience, objects are appearances. By abstracting the sensory aspects from any experience of objects and then separating off everything that belongs to the content of sensation, i.e. empirical intuitions, Kant claims to arrive at a priori formal intuitions purified of all empirical content. Pure intuition is what remains when all that belongs to sensation and all that belongs to thought is removed. Pure intuitions are the forms of appearances, and time is a pure form of intuition.

Time, then, is not derived from experience, but presupposed to it. The succession and coexistence of things in time presuppose time as a form that underlies all possible sense experience a priori. Time can be thought to be empty, but so long as anything is given in experience its temporal aspect cannot be eliminated.[5] Time, then, universally conditions all possible intuitions and so all possible appearances. While it is contingent that there exists any empirical consciousness, time is necessary relative to the actual experience of objects; we could not experience objects atemporally. Time cannot exist independently from experience; nor can time be inherent in things, as they are in themselves independent of their capacity to be known. As an a priori condition of the very possibility of objects appearing to us, time cannot be inherent in the objects it conditions, at least insofar as they are unconstrained by the conditions under which they can be intuited. Appearances only exist because of our interaction with things as they are in themselves. Without the conditions that make that interaction possible there could be no appearances. Since time is a necessary condition for any possible sense experience, Kant asserts that time is nothing with respect to things as they are in themselves.[6] We can know the content of sensation via empirical intuitions, we can know time a priori as a pure intuition, but we cannot know things as they are in themselves.

In spite of this, appearances are neither illusory nor unreal. The natural world does not merely seem to exist outside of us. When we experience objects we experience the appearances of independently existing things. Appearances of objects should not be thought of as subjective. Our experiences and judgments about objects have objective validity, and those objects are objectively real. Time, however, has a double status: it is empirically real and transcendentally ideal. With respect to appearances, time is real and judgments about time have objective validity. Yet with respect to things as they are apart from the conditions of possible experience, time is merely an ideal condition of their possible cognition. Kant will argue that should time be a thing in itself all appearances would be mere illusions.[7] We would not be able to distinguish empirically real objects from our subjective states. Moreover, since divine knowledge would be entirely immediate and so intuitive, were time a thing in itself God would be subject to temporal determinations, as Newton thought. But since God is infinite, Kant argues, divine intuition cannot depend on time, since that would be a limitation.[8]

Were time a thing independent of the conditions of possible experience there would be no way for us to know things about time that we actually know. In particular, Kant claims that we know truths about time that are a priori in that they are strictly universal and necessary. Yet these truths are synthetic, in that knowledge of them requires that we be given intuitions. For their truth cannot be known merely by an analysis of concepts. Kant's examples concern space, but it is likely that, regarding time, he has in mind judgments such as that the order of time is successive. While it is conceivable that time might not be successive, it cannot be true both that time fails to be successive and that experience of objects is as it actually is. Such judgments about time cannot be derived empirically, because they apply to experience universally and with necessity. Nor can they be known through an a priori analysis of concepts, because they require a reference to possible experience. In Kant's terms, we must have an intuition of time, and this intuition must be given a priori because empirical intuitions cannot be necessary or have universal application.

Empirical intuitions – for us, sensations – provide the content of appearances, while time provides their form. As a form, time supplies a single order of succession. This encourages our tendency to represent time on analogy with a spatial line of one dimension. Kant, however, tends to resist this analogy, preferring descriptions of time as having the form of an arithmetic series, because time orders appearances successively while the parts of a line exist simultaneously.[9] Time is a formal condition of any possible appearance, in the sense that it determines the order of the multiplicity of appearances according to relations of succession and coexistence. Because it constitutes a formal condition ordering manifold contents of appearances, time is relational. Yet, since it is an a priori condition of all possible appearances, time is also absolute. The content of appearance is given empirically, but the formal condition of that content is grounded universally and necessarily in time. Time consists wholly in formal relations, but bears on appearances as an absolute condition.

Kant recaptures something of the Greek notion of appearances. Initially, he contrasts time as the form of inner sense with space as the form of outer sense. Through outer sense we intuit objects outside of ourselves. Through inner sense we intuit the flux of our subjective states. Because time is empirically real but not transcendentally real, inner knowledge offers no special self-evidence. Our knowledge of ourselves is empirical and contingent, on a par with our knowledge of the external world.[10] Inner and outer sense both give us appearances which constitute the medium of change. Even if all perception of external objects were illusory there would still be real changes in our subjective states. We know ourselves only as appearances, since, for Kant, change presupposes time, and time is empirically real. In perceiving external objects, we are always aware of both the objects and our perceptual state. Consequently, despite assigning time to inner sense, Kant conceives of time as the formal condition of all appearances, immediately of inner sense and of outer sense as mediated by our perceptual states.[11] The a priori form of time, then, constrains all intuitions and so all appearances.

Again we must emphasize the reversal of the traditional priority of change and time. Kant concludes that time constitutes an a priori condition for the appearance of change and motion. Since time is only an a priori form of possible experience, of appearances, and is presupposed to any possible change, change and motion can occur only in appearances. Kant argues that change would be incomprehensible if time were not already given, because change requires that one and the same thing take on contradictory qualities.[12] This can make sense only if things take on such qualities at different times. Change presupposes time because time orders appearances according to relations of succession and coexistence. Contradictory qualities can, then, succeed one another without incoherence. In this way, for instance, an object can both be and not be in one and the same place, and so time makes motion possible. Therefore Kant sides with Newton against Aristotle, in that time is a prerequisite of motion, not an aspect of it. Not only does time not derive from change, but perception of time does not require any perception of change. For time itself is given in intuition as a condition of all perceptions of change. Moreover, while things may change in time, time itself does not change.

THE FUNCTION OF TIME

Time orders all appearances in a single order of succession. This formal order unifies the multiplicity of appearances in a single encompassing time. For experience to be possible the multiplicity of appearances must be unified. Time is a unity of the multiplicity of appearances. However, time itself is also manifold; it can be divided into an indefinite number of times. These times derive from limitations placed on the unique unitary time.[13] Different times, such as the length of the day, then, consist in parts cut from the one time. The one time is continuous and infinite. Time is infinite in that it is unlimited; given any time interval, a greater time interval exists that includes that interval. Determinate magnitudes of time are possible only through limitations on the one time. Time is given a priori in intuition as a totality. For it can be divided into parts only by an analytic division of the given whole. Time must, then, be understood as a unified infinite span, and so fundamentally as an unlimited, unified order of succession.[14]

The actual infinity of time shows that time is given in an intuition, not through a concept. Kant claims that no concept can itself contain an infinite number of parts, even though it may refer to an infinite number of things. Concepts, by their nature, must refer to many possible instances. Cognition of particular unities, on the other hand, requires an intuition. Only an intuition can contain an infinity of parts within it. The one time, then, cannot be composed from its constituent parts; it must, since it is singular and contains an infinite number of parts, be given in an intuition. Hence time must be given as an unlimited unity prior to and encompassing all appearances. Moreover, Kant argues, that judgments like "different times cannot be simultaneous" are synthetic and as such cannot be derived from concepts alone.[15] So time is not only the a priori form in which all other intuitions must be given to cognition; it is also itself given in an a priori intuition.

While Kant thinks that time is an a priori condition of all possible appearances and that time is given in intuition, he denies that there can be any knowledge, including knowledge of time, through intuitions alone. Knowledge of objects requires both intuitions and concepts.[16] Intuitions can present objects immediately, but objects can be experienced and known only when those intuitions are represented through concepts. Since concepts unite different appearances of the same object, they can hold together the variety of intuitions according to a rule of combination. Concepts, in turn, combine in judgments that can truly or falsely represent objects. Actual experience never consists in the bare data given in intuition, but always in a rich knowledge of objects and their qualities conceptualized in perceptual judgments. When connected, true perceptual judgments constitute actual experience.[17] The separation of intuitions from concepts analyzes this rich knowledge into the respective contributions of the given object and the conditions of its representation. For time to be experienced and known, then, requires that time be represented through concepts.

However, since it is given in a pure a priori intuition and acts as the formal condition of all possible intuitions, time plays a special role in any possible cognition of objects. In addition to the applicability of time to all appearances, the role time plays in making experience possible engenders Kant's privileging of time over motion. Kant argues that experience is possible only if some concepts are able to relate a priori to objects. Empirical concepts can be shown to be objectively real, i.e. to apply meaningfully to real objects, by appealing to experience of those objects. However, some concepts, the categories, i.e. concepts of an object in general, can apply a priori to objects and so cannot be shown to be objectively real by an appeal to experience. In the crucial argument of the *Critique of Pure Reason*, the transcendental deduction of the categories, Kant tries to explain how it is possible for concepts to apply a priori to objects. He contends that experience is possible only if the categories can apply a priori to empirical objects.

In addition to being given an intuition of time, we have a concept of time. This concept must apply to time because time is a formal condition necessary before any experience of objects is possible. However, unlike time, the categories do not represent conditions of possible intuition. For it is logically possible for objects to appear without having to fall under the categories. Appearances might not accord with the rules supplied by the categories. For instance, the category of causality cannot be derived from experience, since it requires of causal relations a strict universality and necessity that cannot be found in sense experience. It is logically possible, then, that the concept of causality not apply to appearances. The concept may be empty of objective content, acting merely as a subjective rule for connecting appearances, extending beyond what is given by the senses. The problem of the transcendental deduction is to justify how the categories as pure a priori concepts can be conditions applying to objects of experience.[18] The deduction must show how these pure a priori concepts can be connected with sensible intuitions of objects. The argument proceeds from the fact that we actually have experiences of objects. Such experience, then, must be possible. The task, then, is to show how it is possible by showing that the categories are necessary conditions of the experience of objects.

Kant claims that it must be possible for my consciousness of objects to be unified in a single consciousness. Otherwise I would be a different self with each representation. Kant identifies this formal unity as the possibility that all my representations can be accompanied by the representation "I think." This one representation combines all my representations so that they are representations belonging to a single consciousness, a transcendental self-consciousness that unifies all consciousness of objects. This transcendental unity provides the purely formal unity necessary for a priori concepts to function as rules of combination. Without this transcendental unifying function, Kant thinks different appearances could not be represented as belonging to the same object. Each appearance might belong to a different consciousness, so that different appearances could never combine to represent the same object. Without this transcendental unifying function, experience would be chaotic and meaningless. Experience, then, is possible only if it is a unified and rule-governed order. The formal unity of self-consciousness cannot be known through experience, because it functions to combine representations, thereby acting as a condition of experience. So the transcendental unifying function supplies the unity that makes it possible for the categories to function as rules of combination.

The categories are a priori concepts of an object in general, i.e. they are rules that determine what it is for something to count as an object that can be represented at all. Consequently, for any object to be represented it must conform to the categories. However, this does not show that it is possible for the categories to apply to the material given in intuition. For experience, as we have it, to be possible, there must be an a priori connection between the categories and what we are given by the senses. Kant tries to establish this connection, in part, by showing that there is a necessary relation between time, the a priori form of all sensory intuition, and the transcendental unifying function of the formal unity of consciousness. If the a priori intuition of time requires the unifying condition supplied by the formal unity of consciousness, then time must be subject to the categories.[19] Moreover, since time places constraints on how intuitions can be unified, experience, as we actually have it, is possible. For if the categories apply to time they apply to objects of the senses in general, since sensible objects must conform to time as the a priori form of all sensible intuition.[20]

The function that links the formal unity of consciousness with time Kant calls "the transcendental imagination."[21] The transcendental imagination combines the multiplicity of intuitions by means of the formal unity of consciousness. This synthesizing activity of the transcendental imagination combines, not merely concepts, but the material of intuition, making it possible for us to know time as given in a priori intuition. The a priori intuition of time constrains the a priori applicability of concepts because it requires that knowledge of time conform to those aspects of time that do not follow from concepts but which must be supplied by intuition. To represent a determinate time interval we must represent not merely the present, but a span of time encompassing an interval that includes some past and future time.[22] The span of time would remain unknowable if the past, present, and future were not unified into an interval. The transcendental imagination provides that unification by reproducing representations of past time, without which earlier times would be lost to consciousness, and producing representations of future time, which are not yet given. These can, then, be unified with the intuition of present time.

This unification must take place from a perspective that can grasp and combine past, present, and future into an interval. The transcendental imagination provides that perspective by applying the formal unity of consciousness to the intuition of time. This atemporal unifying function synthesizes the moments of the time span and forms the concept of time in accord with the demands of the intuition of time.[23] Time, as given in intuition, does not necessitate that its multiplicity fall under the perspective of a single consciousness. That requirement is imposed by the possibility of knowing time through concepts and is a function of the transcendental formal unity of consciousness presupposed to all judgment. This unity operates through the unifying functions of judgments, the categories. In this way, time in general must conform to the categories for it to be known. So, in addition to being constrained by the intuition of time as the form of all appearances, the formal unity of consciousness, through the activity of the transcendental imagination, constrains time insofar as it is knowable. This mutual constraint justifies the application of the categories to time.

Though time and all that is contained in time must conform to the categories, this is not yet sufficient to yield actual knowledge of objects. For objects to be known they must be given to us through sensory intuitions, and these, in turn, through the purely formal intuition of time. Since knowledge of objects requires that they be given through the intuition of time, the categories apply only to objects of possible experience.

STRUCTURING THE TEMPORAL WORLD

The transcendental deduction only shows that the categories apply to time a priori; Kant must also explain *how* they apply. Appearances can be given in time only through sensory intuitions and in accord with the categories. The categories can legitimately apply only to possible experience given in time. Each of the categories determines a structure through which a thing can appear in time. For a concept to apply meaningfully to an object, Kant thinks there must be something the same, homogeneous, in concept and object. But the categories are wholly heterogeneous from all intuitions, because there are no sensible features of objects that correspond to categories such as substance, cause, possibility, or negation. There must be a priori sensible conditions that mediate between the categories and appearances. These sensible conditions that restrict the application of the categories to appearances also make their application possible. Kant calls them "transcendental schemata."[24] The transcendental imagination links each category to time through a schema. Schemata, then, must contain, a priori, both sensible and conceptual aspects. Schemata are pure sensible concepts that provide rules for the synthesis of the categories and time. Schemata make the categories homogeneous with time because they rest on a priori rules that determine the structure of time transcendentally. Furthermore, time is homogeneous with all appearances since every experience has temporal determinations. The application of the categories to appearances, then, is possible through the formation of schemata mediating between concepts and appearances and determining the structure of appearances in time.

Schemata are not images consisting in particular sensory presentations. Each schema is a sensible concept structured by a rule of unity supplied by one of the categories, together with the connection with sensation supplied by the a priori intuition of time. Schemata, then, are a priori determinations of time in accord with categorical rules. As connections between a priori concepts and the pure intuition of time, schemata dispose objects according to those time determinations. Regarding the categories of quantity, the schemata order objects according to a successive time series of their perception. This applies a priori concepts of number and magnitude to objects in time in the form of a temporal sequence. The schemata of the categories of quality organize the content of time in accord with whether sensations are in time, not in time, or occupy some degree of time. The logical modalities apply to time with respect to how objects occupy time. For an object to be actually in time is for it to exist at some time. For an object to be necessarily in time is for it to exist at all times. For an object to be possibly in time is for it to exist at one time or another. Finally, the schemata of the concepts of relation apply rules of duration, succession, and coexistence, not regarding the order of the perception of objects, but regarding their real connections. These schemata organize and represent the objective temporal order.

The schemata of relation, then, have a special importance, since they dispose temporal objects in an objective order of substances and causal relations. These schemata make a priori judgments about objective time relations possible by providing temporal analogs of the rules embodied in the categories.[25] Experience of objects is possible only through a representation of their necessary connections. Yet the temporal analogs of these necessary connections concern only the existence of appearances in their relation to one another. Existence and its content also depend on empirical intuitions. So the application of the schemata in judgments supplies a temporal rule that regulates appearances without determining their content. These a priori judgments dictate what may arise in the objective temporal order, but not what actually does arise in it. The schemata of substance and cause, in particular, supply necessary and universal rules for the unification of appearances, and thereby function as conditions of empirical relations of duration, succession, and coexistence.

The schema of substance guarantees that contrary qualities can inhere in something enduring at different times. This schema, as the a priori possibility of an object enduring over time, makes true judgments concerning change possible. Every change, Kant argues, must be conceived as an alteration of an enduring substance.[26] Moreover, he thinks, really permanent things can provide a necessary condition for the unification of all objects and events in time, and so in one experience. This means that only in relation to something permanent can there be determinate relations of duration, succession, and coexistence. Since time is the unified form of all appearances, every empirical change must occur in this single temporal framework. Hence, succession and coexistence of objects can be represented only through the a priori form of time. Change, however, can only be known empirically, while time, since it is a form of representing objects and not an object itself, cannot be perceived. Time is given through a priori intuition, not perception. The temporal relations of objects, then, must be known from appearances and the rules that combine them. Consequently, knowledge of succession and coexistence requires that appearances be governed by the a priori rule that change occurs in enduring substances. The schema of substance supplies this rule as an application of the category of substance to time as the sensible concept of something enduring over time.

Changes occur through the succession of states in enduring substances. This successive ordering of states must then inhere in an objective temporal order in which one state of an object succeeds another. This order belongs to the objects of perception, and so must be distinguished from the subjective order of the perceptions themselves. The objective order of succession cannot be accounted for by a Newtonian notion of time. In Kant's terms, Newtonian absolute time is transcendentally real. The problem Kant sees in this is that we could not possibly know a temporal order that transcended all possible experience. A transcendent order of this sort would be inconceivable, not merely unknown. Newton's strategy of approximating an absolute measure of time by correcting appearances in relation to other appearances[27] cannot succeed. It illicitly presumes that such an ideal order is conceivable, then merely tries to show how we approximate it through perception. If only appearances are given, the absolute temporal order is not given and any approximations are still merely relative. So absolute time remains inconceivable.

Instead of an incoherent positing, Kant attempts to explain how the objective temporal order can be distinguished from a subjective order within appearances. The objective order of succession is the order of causal relations. The schema of causality brings objects under a rule of succession in time. The schema represents real objects as necessarily succeeding one another in time, governed by a rule of succession. This rule imposes the requirement that every change must have a cause preceding it in time. As in the case of substance, the schema applies the a priori rule provided by the category of causation to the a priori form of time. Here, the category of causation derives its rule from the logical relation of hypothetical connection of judgments as ground and consequent. This type of logical connection between judgments, when schematized by its application to time, becomes the temporal order in which every event must be conditioned by a temporally preceding cause. Because the succession of states of any object must accord with this a priori rule, that succession is objectively valid. This rule shows how it is possible within appearances to conceive of an irreversible temporal order of objects different from the temporal order of subjective states. As conditions of the possible experience of an objective succession, the schema of causality, then, is a condition of the possibility of experiencing an objective order of succession.

TEMPORAL LIMITATIONS

This shows how Kant reinforces the priority of time over motion and change begun by Bonet and Barrow.[28] Time is a condition of substance and causality, number and magnitude, and even real, i.e. empirical, existence. In each case the relevant schema applies to time a rule supplied by one of the categories. Any possible empirical object, since it can only appear in time, must conform to the temporally schematized rules of the categories. Accordingly, all change must take place in time. The schema of possibility gives the rule that contrary qualities cannot inhere in an object at the same time. This provides one aspect of change, the schema of causality another. Every change must be governed by a causal law. The a priori form of time, then, is presupposed to all natural processes. Kant defines nature as the connection of appearances according to laws. Since the schemata supply necessary and universal rules to all appearances, they unify the whole of nature through the connection of time with the categories. Moreover, since magnitude and number, as they apply to nature, must conform to schemata of quantity, time precedes all measures of appearances. Time, then, cannot be the number or magnitude of motion as conceived by Aristotle.[29]

The definition of nature as the law-governed connection of appearances does not commit Kant to an account of nature as an actually existing totality. He argues that the world is neither temporally finite nor temporally infinite, but that the constitution of reason may lead us to antinomies, i.e. valid arguments leading to seemingly contradictory conclusions. Reason demands that any series of conditions form an absolute totality; i.e. given something conditioned, this presupposes the entire sum of its conditions up to and including something unconditioned governing the whole of the series. If the conditioned and its conditions were things-in-them-selves, regardless of whether they can be known to us, then, given the conditioned, all the conditions including the unconditioned would also have to be actually given. When applied to the combination of appearances, this demand of reason generates the idea of an actual totality of nature. Applied to time, the absolute totality of the series of all conditions of some conditioned thing refers to the whole of the time preceding that thing. The world, it seems, must either extend infinitely into the past or have an absolute beginning in time.

These conclusions about the absolute completeness of time supply rules that extend beyond the limits of possible experience. Yet actual appearances contradict the idea of an actual totality of nature. The idea of a sensible totality is problematic. If the conditioned and its conditions are appearances, then, even given the conditioned, the absolute totality of the series of conditions cannot be given. Knowledge of an unconditioned limit can, at best, serve as an ideal to spur further investigation. In experience, Kant claims, appearances must be combined successively. Perception of appearances may regress from any given conditioned thing to its condition, but there can be no experience of an absolute limit to this regress. The only conditions we can encounter in an empirical regress must themselves be empirically conditioned; nor can we perceive a limitation of appearances by nothing. Moreover, there can be no unconditioned element in the series to which all other members are subordinate. Hence the world has no first beginning in time, in which the world would be limited by an extent of empty time.

Objects of experience are always given as conditioned, but the unconditioned is never experienced. Supposing each member of the series were conditioned, we could not know anything unconditioned outside the series that might condition it as a whole. The concept of an absolute totality always exceeds the bounds of our experience of the world. Since we are finite beings we cannot know an actually infinite world. We can know the world's infinity only in the form of an unlimited expansion of the temporal magnitude of the appearances we are given.[30] The magnitude of the empirical regress cannot determine the magnitude of the whole of appearances, since that would anticipate more members of the regress than any empirical combination could encompass. We cannot know anything regarding the magnitude of the world; we can only conceive an indefinite regress of conditions. The world, then, is not infinite in time; nor does the regress terminate at some limit. The regress of past times continues indefinitely: the sensible world has no absolute temporal magnitude.

Although the ideal of an absolutely infinite world cannot be known through experience of the world as it is, Kant thinks that unlimited temporal subsistence is a presupposition of practical reason. The moral obligation to bring about the highest good supposes that this good can be an objective reality. That reality could only come about if our wills could act over a duration sufficient for the fulfillment of the moral law. But the realization of the law would require an actually infinite time, because only by comparison to an actually infinite time could actual particular moral wrongs become insignificant.[31]

In contrast to this practical presupposition, transcendental idealism refuses to accept the concepts of reason by themselves as sources of the knowledge of objects. The non-sensible causes of sensory representations must remain completely unknown to us; they cannot be intuited as objects. Such a cause would have to be a purely intelligible object functioning only as the formal limit of the whole extent of our possible perceptions. Appearances conform to this object, but only as conforming to the rules of the unity of experience. Kant calls such an object "noumenal," and recognizes its problematic nature. Negatively, the noumenal merely indicates the limits of possible experience. In this sense, noumenal objects are strictly unintelligible, transcending all possible experience. The noumenal world is only a formal notion of something indeterminate abstracted from whatever is given to us through the senses. However, Kant recognizes that other beings may have forms of intuition that are intellectual, not merely sensory. If some being has intuitions of this sort it may have knowledge of things as they are in themselves. This would be positive knowledge of a noumenal world. But while we can recognize that other forms of intuition may be possible, all we can intelligibly know is the world as it appears to us through sensible intuitions as organized and combined by the categories.[32]

In sum, Kant assigns time a central role in philosophy. The phenomenal world, unlike the noumenal, can be known through things appearing to us in experience. According to transcendental idealism, appearances in time cannot be things-in-themselves. For time is not one more thing in the world, but stands as a formal boundary condition on phenomena. We cannot know things as they are in themselves because they cannot be given to our sensibility, and time is the a priori formal condition of all possible sense experiences. Transcendental idealism admits the reality of both objects of outer and inner sense, and secures the empirical objectivity of those objects. It does not use the inner as the criterion of reality or deny the existence of empirical objects outside us. Prior to perceiving an appearance we can ascribe reality to it only as given in a possible future perception. The formal intuition of time combines sensory intuitions, determining objects so that they fall under laws that unify experience, laws that originate in the pure concepts of reason when temporally schematized. So the objects of experience are never given in themselves, but only through experience, in time. They have no existence outside temporal appearances.

CHAPTER VIII

Being and becoming

The world ceaselessly begins and ends; every moment it is at its beginning and at its end; there has never been any other world, and it will never be otherwise.

Denis Diderot

Kant's treatment of time focused on its function in structuring the appearance of nature. This approach de-emphasizes difficulties concerning the dynamic aspect of time, temporal becoming. Kant recognized such difficulties in his reluctance to represent time by a geometric line. Yet he still thought that time itself does not change.[1] He saw clearly that temporal moments succeed one another, but he did not acknowledge the internal dynamism of these moments or examine the dynamic aspects of temporal succession. Recognition of the importance of temporal passage arises with criticism of the Kantian distinction between noumena and phenomena.

DIALECTICAL MOVEMENT: HEGEL

While both his treatment of time and his more general philosophical orientation owe much to Kant, G. W. F. Hegel (1770–1831) abhors the notion of the noumenal, either as a limitation on reason or as those aspects of reality available only to a form of intuition we do not possess.[2] Hegel holds that reality is rational and that reason manifests itself concretely as the phenomenal world. To maintain the coherence of this tenet, Hegel must find a way to explain how the rationality of phenomena can be reconciled with what Kant thought of as reason's unavoidable tendency to pose problems for itself that it cannot solve. Kant dealt with this recalcitrant predilection of reason by arguing that it lured rational thought into dialectical illusions, errors due to the use of reason as an instrument for the production of objective truths. Reason seeks the absolutely unconditioned totality of things.[3] However, Kant claimed that when reason attempts to grasp this totality of all conditions in an unconditioned unity it exceeds the bounds of possible experience. Since he thought that objective knowledge required intuition as well as concepts, pure reason without intuitions would illicitly posit objective realities transcending possible experience. Kant claims, in particular, that the antinomies of cosmological concepts must be resolved by treating the conditioned as objectively accessible in phenomena, but also by treating any unconditioned totality of conditions as noumenal, beyond the limits of possible experience.

In response, Hegel finds casting the unconditioned into the noumenal unacceptable. The unconditioned must be given phenomenally. This is possible, according to Hegel, because, rather than an illusion of reason, dialectic itself is phenomenally real. Transcendental logic, which contains necessary conceptual structures of experience, becomes an ontology of being, and being appears as it really is.[4] Kant failed to recognize the contradictory nature of phenomena. He recognized the contradictions arising in the antinomies, but tried to avoid them as subjective though unavoidable errors due to a failure to distinguish indefinitely extended phenomenal series from noumenal totalities. Hegel affirms the contradictory nature of the antinomic propositions. The antinomy of the beginning of time, for instance, amounts to "two simple opposite assertions: (1) there is a limit, and (2) the limit must be transcended."[5] Hegel also agrees that reason strives for an unconditioned totality, but denies that this totality transcends phenomenal reality. Since the real is rational there must be a conception adequate to the unconditioned totality of phenomena. Any concept that pretends to such adequacy but falls short is antinomic. Yet, as Kant demonstrated, such concepts are indispensable. Reason, then, must both preserve and unify the contradictory determinations of its universal concepts. It does this by subsuming these contradictory determinations under a more inclusive rational structure. Hegel's formula for this dialectical rationality regards the absolute totality of phenomena as the identity of identity and difference.[6]

Rather than as an illusion of subjective reason, Hegel sees dialectical oppositions as ineliminable, though partial, realities. He makes sense of contradictions in reality by treating them dynamically.

Immanent dialectical oppositions appear as the restless movement of becoming. The phenomenal world moves and differentiates itself into abstract moments that arise and pass away. The unconditioned totality of appearances "is the arising and passing away that does not itself arise and pass away."[7] The only permanently static aspect of appearances is the dialectical structure itself, which, in Hegel's terms, is the abiding of the vanishing of its abstract moments. Determinate realizations of phenomena do not persist. Instead, passage itself does not pass away, but remains preserved as the totality of pure being, abstracted from any content.

Dialectic logic, then, begins with pure unconditioned being, actually present as the unmediated and undifferentiated determination of the totality of phenomena. Being offers a conception of totality because anything must have being, so nothing can be external to being. Yet because this conception of absolute reality lacks articulate content it amounts to an abstract empty form. As such, being is indistinguishable from an abstract negation of all content. Pure being, then, is the same as nothing. For nothing is absence of content, undifferentiated emptiness. Moreover, since this dialectic takes place among conceptual structures necessary for the appearance of actual phenomena, it assumes the identity of thought and being. Whatever can be thought or spoken of, then, has being. So, Hegel argues, since nothing, itself, can be an object of thought it has being. Pure unconditioned being and contentless nothingness, which at first absolutely exclude one another, are, as a result, also identical. The proposition "being is identical with nothing" denies the difference between these concepts. Yet, Hegel claims, all such statements of identity show the non-identity of their terms because their very statement separates them. Being and nothing must be maintained in their difference and yet unified in their identity. This antinomy, Hegel claims, results from holding on to a strict dichotomy of being and nothing. Neither being nor nothing can subsist by themselves; there must be a third concept, becoming, in which they subsist.

Becoming is the determinate unity of being and nothing. Being and nothing vanish into one another; each passes over into the other. Being is not the same as nothing. Both are preserved in the unity of becoming, but only as vanishing determinate moments. The distinction between being and nothing is abstract; neither can subsist on its own. Being and nothing can subsist only in transition into one another. The restless movement of becoming constitutes their unity or subsistence, a movement that asserts their identity, while preserving their difference. Becoming unifies their difference, for all phenomena appear "in an intermediary state between being and nothing."[8] Becoming distinguishes being from nothing in a dynamic equilibrium of coming to be and passing away in which being and nothing continuously separate and reunite. Moreover, being and nothing each contain a unity of being and nothing. Coming to be passes from nothing to being, and passing away passes from being to nothing.

The dialectic of becoming displays the nature of dialectic movement itself. Phenomena form an unconditioned closed totality through the infinite circular figure of perpetual becoming oscillating between being and nothing. The transition of being and nothing cannot be conditioned, since any further conditions must express determinations less abstract than being and nothing. All further logical determinations of phenomena exemplify becoming.[9] The simple identity of being and nothing does not remain, because phenomena must be conceived in terms of more determinate and concrete content. Indeterminate being inherently produces contradiction and so transition, while determinate content does not relate purely to itself, but makes sense only in contrast with other determinate contents. Time, for Hegel, is nearly identical with this substantive dialectical movement.

Time is becoming directly intuited as an aspect of nature.[10] Time is not a container in which everything comes to be and passes away. It is an abstract moment of the actual continuous coming to be and passing away of nature. Motion and change in the natural world presuppose the dynamic structure of becoming. The natural world comes to be and passes away through the appearing of the dialectic movement of becoming in the unity of past, present, and future. Being unites with non-being in each of these moments of time. The past is the actual being of the world that has ceased to be; the future, the non-being of the world that actually will be. The present unifies the past and future negatively, since the present comes to be through the past ceasing to be, yet the present exists now because its non-being lies in the future. Time is the contradiction in and ceaseless motion of finite beings, which fall short of the totality of phenomena. Universal natural laws express the dynamic equilibrium in this motion. Such laws govern natural processes, reconciling differences, but do not themselves undergo such processes.

While Hegel accepts Kant's view that nature is given through intuition, he also holds that there can be no source of intuition beyond the phenomenal world. Reflective consciousness turns intuition back on itself so that self-consciousness gives itself its own objects. For Kant, the unity of phenomena was a function of self-consciousness. Hegel recognizes this constitutive function of self-consciousness, but since he identifies reality with the absolute totality of phenomena self-consciousness becomes a condition of the substance of reality itself, not merely of knowledge. The unifying function of self-consciousness not only rises above dogmatism and skepticism, but also overcomes the diremption of subject and object. Moreover, since time is the pure form of intuition, time is the form of pure self-consciousness. Time is becoming given in every intuition of nature. But, the substance of appearances is subjectivity. So, while time appears in nature, it is also the becoming of self-consciousness.

Becoming, however, has many empirical shapes. It appears in nature, the processes of life, self-consciousness, ideology, intellect, social life, history, religion, and philosophy. These phenomena arise and pass away, since they are only partial determinations of the whole. The flux of becoming only achieves stability with the absolute subjectivity of the phenomenal world as a whole. Hegel calls this soul of the world *Geist* (Spirit). Spirit is the self-conscious concrete substance uniting all that appears. Spirit is the stable absolute self in which the flux of becoming occurs, articulating the restless dialectical movement of all partial realities. The unity of self-consciousness is the expression of Spirit realizing itself. All phenomena are expressions of this self-consciousness, of a self-positing Spirit equivalent to the whole of reality. Events of history, for instance, are concrete expressions of Spirit.[11] Time, then, includes the becoming of the ontological and historical self-genesis of Spirit.

The temporal becoming of Spirit expresses itself in its most immediate form in the now. At first it seems as though the now is apprehended through the senses in an immediate connection between consciousness and its object. The now seems to affirm being and negate the abstract and absent moments of the past and future. However, like Aristotle, Hegel denies that time can be constituted by such a simple series of nows.[12] But, unlike Aristotle, Hegel explains temporal continuity by the unity of self-conscious Spirit. For were time a series of nows we could not experience the unity of phenomena. Each now would constitute an isolated phenomenon. Nor could either the particular object or consciousness alone constitute the now. For the now may remain the same while its phenomenal content changes, and that content may remain the same through different nows. As an isolated particular, the now can only be an abstraction from the dynamic unity of phenomena in which subject and object are each mediated by the other. Since it can indifferently have or not have a particular content, as an actually given point in time, the now is a universal in perpetual dialectical movement. The abstract now is an absolute end of the past or beginning of the future, while the universal now is a flowing limit. Each concrete moment of time exhibits an internal dynamism.

Hegel presents the reality and permanence of becoming as the totality and unity of determinate oppositions.[13] Time is the coming to be and passing away of the Absolute Spirit. The evanescence of Spirit is essential, since that flux itself is always actual and does not pass away. The totality of appearances is an absolute flux, a universal differentiation. Every finite objective content is a temporal appearance, dissolving as its differentiation disappears when reflected back into itself by its opposite. Rational differences and the rhythms of their separation and dissolution comprise the organic whole of Spirit's content. The life of Spirit is the independent and enduring resolution of this infinite motion, the supersession of all distinctions. This infinite being of Spirit is no longer empty abstract being, but encompasses the substantial pure flux within being itself. "This infinity . . . is itself every difference, as also their supersession; it pulsates within itself but does not move, inwardly vibrates, yet is at rest."[14]

This flux is the essence of time. Anything that appears as a conditioned articulation within Spirit is subject to this dialectical flux of time. Spirit, as fully concrete and articulated reality, contains only temporal articulations. Rational distinctions, and the partial realities that exemplify them, are ineluctably temporal. The phenomenal world, then, is temporal through and through, and the dialectic has the shape of a pure becoming. But Spirit itself is not subject to temporal flux any more than was the empty concept of being. Spirit is absolutely present. It does not abide and vanish, since it exists as the completely rational whole of reality, which lacks nothing. Spirit is the unconditioned reality superseding temporal flux as its stable equilibrium, the permanence of the infinite circular movement from undifferentiated content to immanent differentiation and back again. This absolute timelessness of Spirit is not duration, but the absence of time. Time, then, forms the boundary condition of phenomena as the limit of the phenomenal world. All phenomena are temporal. There are no boundaries to phenomena in the sense that there is nothing outside phenomena. The only thing beyond time is its unity in Spirit. Spirit expresses itself as the articulation of the unconditioned rational totality of things. The phenomenal world is always conditioned by time. In this sense, time itself is eternal, since everything must appear in perpetual becoming.

Spirit, for Hegel, is a general whole concretely inclusive of all phenomena; it is not merely an abstract being, but a concrete universal. Yet it shares with abstract being the quality of the negative. Whereas abstract being is equivalent to nothing, concrete universal Spirit achieves its universality precisely by negating the nothingness of abstract being. Universal Spirit can be realized in the concrete totality of phenomena because it negates the oppositions between partial beings and reconciles these articulations in the equilibrium of a higher union. This reconciliation not only negates these oppositions, but appropriates their content as always already contained in the eternal being of Spirit. Nothing occurs among phenomena except what unfolds from what is already present in Absolute Spirit. Without this unchangeable, and so timeless, co-presence of all differences, Absolute Spirit would diffuse into the pure immanence and succession of its moments. It is precisely this diffusion that Nietzsche will seek to affirm, while putting aside any recourse to universal Spirit or the labor of the negative.

NIETZSCHE CONTRA HEGEL

Friedrich Nietzsche (1844–1900) diverges from Hegel on three points that, together, propel him toward a radically different model of time.

First, Nietzsche displays the nihilistic consequences of an eternal universal being. Hegel's Absolute Spirit was incommensurable with time. Spirit posited time only in order to unfold the permutations of what it already contained. Spirit achieved universality by negating differences and so negating becoming. Absolute being served as a standard by which phenomena were judged. All becoming was condemned to relative unreality by falling short of this standard of being. Time and becoming could be redeemed, for Hegel, only by the universality of the absolute. Nietzsche attempts, instead, a revaluation of becoming.[15] He rejects all transcendent being. There is no universal higher presence of Spirit nor any absolute whole that binds and determines an order in the multiplicity of becoming.[16] Only within becoming do identities, equalities, and equilibria get constituted. Lacking being, the separation of appearances from their ground collapses. Lacking the standard of being, becoming is innocent; it cannot be condemned.[17] Needing no redemption, time disperses into the multiplicity of becoming, no longer mere partial realities or images to be superseded by an inclusive whole. There is nothing beyond becoming and multiplicity; neither is illusory.[18]

Second, Nietzsche exposes the negative evaluation embedded in dialectical movements reappropriating partial realities, which, for Hegel, comprised temporal flux. On Hegel's account, time is determined by the structure of reflection – a dialectical succession of partial states propelled by partial negation, to be completed in the concrete universal. The dialectical interpretation of becoming reproduces the judgment of Anaximander: becoming is unjust. Limited things expiate their guilt in mutual struggle. And the totality of things derives from an original unlimited being that, as becoming's unchanging origin, redeems it.[19] The dialectic labor of the negative condemns becoming to nothingness: first as the negation of the oppositions between partial realities, and then as nothingness due to the reappropriation of those oppositions by a higher unity. The dialectic requires two negations to affirm being.

Instead of this negative evaluation of becoming, Nietzsche discovers an affirmation of becoming and differentiation.[20] Without a higher unity, becoming can be affirmed in itself. The contradictions of the dialectic misinterpret differentiation. No longer defined by opposition, appearances express positive forces. While opposition relates abstract products, differentiation is a positive principle of production. Each thing is the product of the interaction of forces, a will-to-power, which also produces the thing's qualities. Change is no longer an unreal abstraction, but is inherent in the affirmation of becoming. Nietzsche's conception of time as eternal return offers a double affirmation of becoming that does not submit to atemporal being.[21]

Finally, Nietzsche displaces the unifying function and privileges of self-consciousness. Spirit, for Hegel, was the self-conscious unity of substance and subjectivity. But with the affirmation of becoming, this unity of the subject disperses into multiplicity.[22] No longer a unity, subjectivity can serve as a condition neither of the unity of phenomena nor of their reality. Having lost its inherent duration, the subject, like the thing-in-itself, devolves into the multiplicity of phenomena: phenomena no longer denigrated to the status of unreal or partial realities. Instead of consciousness as a standard and condition of phenomena, other forces produce consciousness in their service.[23] Consciousness, then, is an effect – an essentially reactive awareness of the effects of the forces that produce it. The affirmation of becoming makes duration, substance and consciousness problematic. Absent a transcendental subject, each must be constituted by yoking together the moments of becoming. The moments of becoming must be synthesized in an occurrence that happens again and again.[24] The possibility of this synthesis without recourse to a transcendental subjectivity orients the problem of time.

NIETZSCHE: ETERNAL RETURN

A rhythmic tension synthesizes the moments of becoming.[25] Time is not a flow, but a pulsation. If time merely flowed it would lack tension and no differentiation would occur. The whole of time is at stake in the rhythmic tension of the moment. This tension arises between the becoming of being, the differentiating and generative element in appearances that Nietzsche calls will-to-power, and the being of becoming, which allows the moment to pass and, in passing, allows differentiation to recur. Nietzsche calls this eternal return. Being comes to be at every moment. Past and future do not exclude one another but emerge together in each moment of becoming.[26] Past and future, and consequently all time, arise in the moment. This moment is not in time as one moment among many in a container; it is time. "Inside," "outside," and "containing" are spatial metaphors that mislead us in understanding time.

As we have seen, time bounds appearances. Since, for Nietzsche, there is nothing beyond appearances, the world itself actually appears. The contrast between appearance and reality dissolves in Nietzsche's empiricism. Appearances do not conceal a transcendental being or a hidden essence. Consequently, time must act as an effective ontological principle.[27] The rhythmic tension of the moment conditions both change and stability.[28] The world consists, then, of forces that struggle for dominance. While forces differ quantitatively, these differences cannot be measured numerically, by means of abstract units. For, all unity must be produced as an effect. Differentiation of quantities, then, determines the quantity of each force.[29]

Appearances comprise the becoming of forces. Every force is related to another force qualitatively, either as commanding or obeying, superior or inferior. Will-to-power is an interpretation of all occurrence, the creative power of becoming.[30] It is an ontological notion and should not be confused with the psychological will. Unlike the Neoplatonic emanation of becoming out of being,[31] will-to-power coincides with an emanation of being out of becoming. Existence is constituted through the production and affirmation of differences. Will-to-power is the principle of synthesis that differentiates quantities of force and interprets their qualities as superior or inferior. With no principle of generation outside of appearances and becoming, will-to-power both produces the relations of forces and is itself manifested in those relations. As productive power, will-to-power is a tendency to increase, for a force to go to its limit vying for domination with other forces. As manifestation, will-to-power is an affect, the capacity of a force to be affected, not being or even becoming, but a pathos or passion.[32]

The qualities of forces are active and reactive.[33] A force is creative and active when it goes to the limit of what it can do, dominating other forces by affirming its difference from them. A force is reactive when it separates an active force from what it can do. This turns the active force against itself, making it into a reactive force. Reactive forces operate by negating what opposes them rather than by affirming their difference. Consciousness, which separates the doer from the deed in its reflective or dialectical structure, and utility, which separates means from ends, are reactive forces in this sense. Reactive forces never form a greater force than active forces. Yet reactive forces may dominate active forces by making an active force become reactive, by separating an active force from what it can do.[34]

Will-to-power interprets, determining the active and reactive qualities of forces. This interpretation is an evaluation which assigns becoming an affirmative or negative value. Nietzsche interprets the western world as suffering historically, psychologically, and metaphysically from a nihilistic evaluation of becoming. Historically, forces which portray themselves as promoting the highest values conceal a negative evaluation of historical becoming. Psychologically, nihilism is the aimlessness experienced when the highest values devalue themselves, when unity, truth, and useful goals all fail to give life meaning. Metaphysically, nihilism denies and condemns becoming in favor of being. Nihilism is a will-to-power that interprets becoming as without value. However, nihilism also makes us conscious of will-to-power. So it allows us to affirm that "This world is the will-to-power – and nothing besides!"[35]

To affirm becoming and depart from nihilism, Nietzsche turns to eternal return. Eternal return overcomes nihilism by taking it to the limit of what it can do. This makes nihilism into an active force and eliminates the reactive forces. Eternal return pushes negation to its limit, beyond all partial negation, to the most extreme nihilism: the endless recurrence of a world without meaning or purpose, and without hope of annihilation.[36] This extremity of nihilism is part of what Nietzsche means when he proclaims that the eternal return is the most difficult thought to bear.

Moreover, eternal return cannot be easily incorporated into our usual conceptions of the universe and of time. In Nietzsche's sense, eternal return denies the possibility that the universe achieves a final state. Eternal return, in this sense, is incompatible with a static world of either being or nothingness. Both teleological and natural or mechanistic conceptions posit a completed or final state of things, albeit in different ways. Mechanistic conceptions of eternal return portray it as a repetition of a completed cycle. Given Newtonian infinite time and a finite quantity of force in the universe, eternal return reduces to the repetition of a sequence of events in time. If a single sequence does not occur there can be no eternal return.[37] Even if such a sequence would occur and recur endlessly, it could not incorporate eternal return, since it posits the repetition of being rather than a universe of becoming. Teleological models posit some goal as a final state by which it is judged whether the world has achieved a state of being. Again, becoming is deprecated in favor of a state of being. While both mechanism and teleology are nihilistic, neither is able to take nihilism to its limit and overcome reactive forces of utility and consciousness. Nihilism is taken to its limit only when becoming is reduced to meaninglessness without recourse to a world of being.[38]

Part of the difficulty comes from treating time as duration. Time can be conceived as duration only if time has being as a sort of extension or by reference to some standard. Eternal return is neither a series of recurring events nor a special static state. Conceived as duration, time cannot return since it is already completed and determined. So eternal return involves no persisting duration and no endless continuance. What endures cannot pass, and so cannot return. Newton required an omnipresent deity to maintain duration. Without a transcendent being, the past and future become constituted in each moment. Nor is duration necessary for time's infinity, because in eternal return the moment inexhaustibly renews itself. Without duration, return happens in each moment; it is the occurrence of time itself.[39]

Time does not flow, but occurs in each moment. Instead of the river of time, Nietzsche's prophet-hero Zarathustra posits the gateway: moment. In the gateway two paths confront one another, coming into the moment from an indefinite past and an indefinite future.[40] The past does not arise out of an initial state; nor does the future attain a final state. The moment brings past and future together in dynamic tension; each moment is eternally already open onto past and future.[41] It synthesizes the moments of becoming because the moment returns of itself. The moment passes and, in passing, it returns.

> And if everything has been there before – what do you think, dwarf, of this moment? Must not this gateway too have been there before? And are not all things knotted together so firmly that this moment draws after it *all* that is to come? Therefore – itself too?[42]

Eternal return is "the closest approximation of a world of becoming to a world of being,"[43] because each moment opens on to all other moments and determines its own relation to them. Temporal passage in eternal return allows no enumeration. The moment is neither one nor many: it unifies all moments, but since it passes it also returns, thereby continually differing from itself. The moment is that which differs from itself and, in differing from itself, returns. Becoming is a unity of multiplicity, manifold but, since it includes itself, uncountable. Were all things not "knotted together so firmly" the moment could not pass.

The moment passes in a cycle of arising and passing away of moments. The cycle of time is not Plato's great year, but rather the dividing moment that arises again and again, incorporating all of time.[44] The cycle does not follow the course of linear time, but constitutes being from within each moment.[45] The moment's returning returns. No moment ends, for each allows another moment to arise and incorporate it. Time is occurrence, a continuous arising, not an independent extension. So, rather than having an abiding place in a series, the series of moments arises and perishes within each moment. Succession is produced and reproduced within each moment. Eternity, then, is neither everlasting nor atemporal. It is the release and distribution of every moment within each moment. Time and eternity do not stand in opposition to one another. The inexhaustible return of the moment, occurrence itself, is eternal.[46] In a world of becoming the rhythmic cycle of arising and passing makes timelessness and transcendence meaningless. Eternity is not transcendence, but sheer occurrence continuously incorporated as the moment returns. The rhythmic pulsation is the form of becoming and the appearing of the world.[47] Rhythms have no end-states or goals. At each moment they are distributed over past and future in a dynamic structure of tension and release that is never fully determined because it will be modified as the moment passes.

Becoming inheres in the passing moment through an unconditioned synthesis of the moment with itself and all other moments. The moment stands in a synthetic relation to itself which determines its relations with every other moment. Eternal return synthesizes the forces that produce diversity; will-to-power is the productive principle of this synthesis.[48] The arising and differentiation of forces in will-to-power occurs in the moment that eternally returns, which then differentiates things anew. Eternal return distributes the differentiation and interpretation produced by will-to-power, i.e. eternal return "creates new laws of motion . . . not new forces."[49] Unlike the Kantian synthesis that unifies appearances, eternal return does not constitute a stable unity through determinate categories. The being of the world is constituted, instead, out of the becoming of the moment. Eternal return deactualizes the moment and its unities.

The moment that eternally returns, then, is not a temporal present. Eternal return ascribes passage to time without recourse to any present being or anything being present. A world of becoming is never complete and nothing in such a world can ultimately be finished or whole. Eternal return distributes the products of will-to-power both particulars – the identities of individual beings – and universals – the possibilities of being. There are no stable identities and no permanent qualities. Identities and qualities are constituted only by their returning again and again. Their totality is always open, their completion always deferred. Will-to-power, the productive tendency to increase, then, is an a priori principle of return. This undermines any metaphysics of being. For in a world of becoming the distinctions and priorities of whole and part, particular and universal break down. The whole/part distinction gets destabilized, since no identity is settled once and for all. The universal/particular distinction remains indeterminate, since new qualities may be generated in each moment. Becoming, the self-inclusion of the moment, then, cannot appear as a being with qualities; rather, beings and their qualities arise out of becoming. This means that being arises out of pure intensities that return on themselves, oscillating so as to divide and reunite with themselves, and consequently take themselves as their own objects.[50] In this way, designations and articulated appearances arise out of becoming. Time, then, is neither an aspect of change nor an absolute extension; it is a constitutive intensity.

Eternal return guarantees the paradoxical nature of the moment. The self-reference of the moment is only possible in its return, which separates it from itself and allows it to point to itself. Such self-reference is both a creative tension and the condition of meaningfully articulated appearances. In differing from itself, becoming produces structures of reflection and appearances. Self-referential intensities become representations, which are themselves occurrences, not merely copies of pre-existent beings.[51] In occurrence, there is no separation of appearance from reality. Occurrences act and have effects producing identities and possibilities.[52] The world is will-to-power, but the world is only related to itself through time. Through eternal return, each moment represents all that will or can ever happen. So eternal return is neither subjective state nor fact of nature, neither arbitrary nor determined. Eternal return, the occurrence of time, then, is a boundary condition on phenomena.

In distributing and interpreting forces, will-to-power produces the occurrence of chance. Eternal return is the being of becoming, the unity of multiplicity, and the necessity of chance.[53] As the being of becoming, it guarantees the passage of the moment. As the unity of multiplicity, it represents all occurrence. As the necessity of chance, it both affirms actual occurrence and necessitates that the occurrence of chance returns. Differentiation returns inexhaustibly. As becoming, which differs from itself, eternal return reproduces and affirms the occurrence of chance itself. In representing all occurrences, eternal return affirms all forces and so affirms the necessity of the play of the whole chance. As pure occurrence, then, eternal return gathers together the chance configuration of forces and unites every event with necessity.[54]

Eternal return, uniting chance and necessity, is fate, not process. It is chance that certain forces coincide at any particular moment, and the unconditioned synthesis of eternal return provides a principle of necessity, a principle of selection. Fate exceeds any particular force, and no force can be separated from the totality of forces. Yet, since nothing can fall outside of eternal return, chance occurrence cannot be constrained by any external forces. Forces themselves necessarily determine one another, and the totality of forces necessarily returns. So all forces that return go to the limit of what they can do. Fate consists in the necessity that impulsive forces coincide and pass through uncountable transfigurations. Hence, fate does not set human activity against implacable forces beyond its control: human willing merely circulates among a multitude of forces. Fate exceeds human will and cannot be constrained by any teleology of being.[55]

Where a metaphysical being would condemn the world of becoming and a human will would react against an irreversible past, will-to-power wills "the eternal joy of becoming."[56] Were reactive forces to return, nihilism could not be overcome. But, since only forces that go to their limits return, reactive forces cannot. Reactive forces do not return because in unifying all force relations eternal return exceeds any given relations of forces and brings back pure occurrence, the occurrence of becoming, multiplicity, and chance. Since each moment includes all that can happen, time is not a relation between appearances and a transcendent ground, an aspect of change, or an empty form. In passing, each moment releases itself and all that occurs within it. Neither identities, equilibria, nor evaluations return, because sameness is constituted within each moment and, in returning, the moment always differs from itself, extending into other moments. Without a transcendent being, final purpose, or absolute totality no questions of justification can apply to becoming. The passing of the moment and its creative return constitute an innocence of becoming.

This creative return of forces does not react against an irreversible past, since return releases will-to-power from being bound by the distribution and interpretation of the moment that has passed. Will-to-power interprets by creating new forms of existence, not by submitting to past forms, nor by applying given values. Will-to-power becomes creative through the selective principle of eternal return. Will-to-power differentiates, while eternal return evaluates and selects that which returns. This ontological selection eliminates reactive forces, since it demands that whatever is willed must also be willed eternally to return. This requires that each force increase as far as it can go. These active and intensive forces both constitute the moment and exceed it. In exceeding the moment, intensities affirm themselves and thereby affirm every moment. The principle of eternal return affirms past and future as constituted in each moment. Each moment revaluates all time. To affirm any moment, then, is to affirm all existence.[57] For affirmation is creation, the release of existence in the generation of new values.

Affirmation diverges from, rather than opposes, negation. Negation, as Hegel made clear, leads to the triumph of being; it reduces becoming to oppositions between reactive forces. Instead, affirmation differentiates; it actualizes the principle of eternal return in selecting active forces.[58] In eliminating reactive forces, eternal return transforms even negation into a mode of affirming. Negation too must go to the limit of what it can do. The limit of negation is the complete elimination of being, and this affirms becoming. For, if being cannot abide, each moment must pass and differ from itself. Not only does eternal return affirm becoming, but it affirms the being of becoming, i.e. it affirms its own affirmative character, by affirming its differing from itself. The passing of the moment in return must incorporate returning itself by representing returning, itself, in its passage. In this way, eternal return affirms itself as its own object. In differing from itself becoming returns, and in incorporating the necessity of its own return becoming returns eternally. This affirmation of becoming affirms that being must arise as pure occurrence. Nothing can come into being without changing its character.[59] Being is the being of becoming; its only character is eternal return.

Part Two

Contemporary traditions

INTRODUCTION TO PART TWO

Thus deeper and deeper into time's endless tunnel, does the winged soul, like a night-hawk, wend her wild way; and finds eternities before and behind; and her last limit is her everlasting beginning.

Herman Melville

Part Two will abandon the linear historical order of Part One. Instead, it will follow three philosophical traditions through the twentieth century: the analytic tradition, which concentrates on McTaggart's problem; the phenomenological tradition, which concentrates on the unity of temporal appearances; and the distaff tradition, which concentrates on temporal synthesis. At the beginning of the century each of these traditions proposes an account of time which, through criticism, matures by the end of the century. Part Two will follow developments in each tradition for two chapters, then return to the beginning of the century to describe the next tradition. (Special terminology used in this introduction will be explained and clarified in the pertinent chapters of Part Two.)

These traditions focus on different problems and relate to one another only tangentially at certain points. Beginning with McTaggart's problem, analytic philosophy critically develops the contrast between static time and temporal becoming. Beginning with Husserl, phenomenology critically develops problems concerning temporal appearances and their unification. Beginning with Bergson, the distaff tradition critically develops the problems of temporal synthesis and the generation of novelty. However, the phenomenological tradition acknowledges and incorporates the aspects of time studied by the analytic tradition, and the distaff tradition acknowledges and incorporates the aspects of time studied by the phenomenological tradition and thereby of the analytic tradition. Since this is the case, we will discuss these traditions in the sequence of greater inclusiveness.

Although the traditions use different methods, have different problems, and emphasize different aspects of time, there are some common themes that cut across these differences. We shall emphasize six themes of importance in all three traditions.

First, each of the three traditions has historical roots in both the ancient and the modern world. McTaggart's problem arises because of the need to deny the metaphysical scheme in Iamblichus' two sorts of time. Aristotle distinguished before and after and tied time to what is countable, presaging B-series accounts; he also investigated the now, presaging A-series accounts. The Epicureans distinguished perceived time from conceived time, and modern thought conceived the distinction between subjective and objective time as well as the mathematical representation of the time line, all of which serves as background to discussions in analytic philosophy. The phenomenological tradition inherits from the ancients variations on an Aristotelian account of natural time, and an Augustinian concern with time and distensions within the subject. It also derives an empiricism from Locke and his successors, combining it with aspects of transcendental philosophy inherited from Kant. The distaff tradition inherits aspects of the Stoics' distinction between bodies and events, Duns Scotus' univocal ontology and his characterization of potentiality by way of the virtual and its actualization, Spinoza's conception of immanent causal expression and causal relations of forces acting in succession, along with Leibniz's recognition of the need for a cause of the whole temporal series, and Nietzsche's account of the eternal return and the production of representation within the moment.

Second, since time is a boundary condition of appearances and all twentieth-century traditions accept the empiricist project of tying all possible reality to appearances, each tradition attempts to understand time from within appearances. However, they differ regarding what constitutes appearances. Although empiricism is an epistemological position, every empiricism must presume an ontology of appearances; the three traditions differ regarding these ontologies of appearances. For the analytic tradition, appearances are particular perceptions from which other realities can be inferred even though they may not be directly observable. For the phenomenological tradition, appearances include not only simple impressions, but also universals, syntactically formed states of affairs, and intentionalities. For the distaff tradition, appearances consist in mixtures of multiple tendencies and the differences, similarities, subjects, and objects those tendencies produce.

Third, in each tradition time is closely connected with existence. With the analytic tradition, A-theorists either treat past, present and future as forms of existential operators, or argue that appearances of A-characteristics exist and are ontologically real aspects of the objective world. B-theorists argue that earlier and later are objectively existent relations. Within the phenomenological tradition the ontological difference between being and beings is a distinction in modes of being that is grounded in different aspects of time. For the distaff tradition, temporal differenciation produces being, so that the virtual actualizes being univocally.

Fourth, each of the traditions privileges a fundamental ontological relation. The analytic tradition privileges the relation of quantifiable particulars and their properties. The phenomenological tradition privileges the nexus of intentionalities, given through what is invariant in examples, and its being-in-the-world. The distaff tradition privileges the relation of virtual to actual, univocal and continuous actualization.

Fifth, the difficulties of time and representation have their roots in the nineteenth century but come into the twentieth century by three different paths. Questions concerning the reality of time, the unity of time, and its determinacy tend to generate problems of representation. In each tradition the relationship of time to its modes of representation becomes an important concern. However, the traditions differ regarding the role, analysis, and impact of the modes of representing time. In the analytic tradition analysis of the sense and reference of different linguistic forms for representing time plays a pivotal role in attempts to resolve McTaggart's problem. In the phenomenological tradition the constitution of a temporal span and the interpretative import of its unity turn on explication of the modes of representing time. In the distaff tradition bodies and causes generate the articulation of temporal becomings, including all representations and their parallel articulations in states of affairs.

Lastly, each of the traditions connects time with pragmatic

concerns. In the analytic tradition A-theory appeals to pragmatic concerns to show the ineliminability of A-characteristics, while B-theory takes this as indicating their subjectivity – as counting against their objective reality. For the phenomenological tradition, pragmatic considerations order and delimit everyday appearances. For the distaff tradition, pragmatic concerns account for the spatialization and abstract quantification of concrete *durée*.

CHAPTER IX

McTaggart's problem

Truth . . . rival of time.

Miguel de Cervantes

The present . . . indivisible and flowing point, on which man
can no more stay fixed than on the point of a needle.

Denis Diderot

In 1908 John McTaggart Ellis McTaggart (1866–1925) published an
article containing an argument purporting to show the unreality of
time.[1] While this argument was later included in McTaggart's major
work,[2] it stands alone in both its form and influence. McTaggart's
ontological position is a variant of Hegel's. Yet his argument for the
unreality of time does not draw on Hegelian concepts, particularly
the notion of a sufficient description of substances, which domi-
nates his other ontological concerns. The argument that time is
unreal stands apart from McTaggart's peculiar ontology, for it
concerns features of time as it is experienced. The importance of
the argument is that it raises questions about the coherence of
temporal appearances. Consequently, debate about the argument
has been inherited by empiricist philosophers, especially analytic
philosophers, who tend to recast the debate in terms of questions
concerning linguistic representations of time.

McTAGGART'S ARGUMENT FOR THE
UNREALITY OF TIME

McTaggart claims that nothing that exists can have the property of
being in time; in this sense, time is unreal. Time, he asserts, is always

experienced as ordered in two distinct series: he calls these the A-series and the B-series. McTaggart defines an event as the content of a temporal position (a moment), and writes of events as well as temporal positions as ordered in these two series. In an A-series each moment or event is characterized as future, present, or past. Every moment or event changes with respect to these A-series characteristics. McTaggart proposes that change with respect to these characteristics constitutes the passage of time. In a B-series moments and events stand in relations: t_n is earlier than t_m and t_m is later than t_n. If one moment or event is earlier than or later than another the positions or events stand in those relations permanently. The A-series demarcates dynamic time, the B-series static time.[3]

McTaggart assumes that if the application of a distinction to reality implies a contradiction that distinction cannot be true of reality. Given this assumption, he argues that neither A-series distinctions nor time can be real. For, first, if time is real, then A-series distinctions must apply to reality and, second, the application of A-series distinctions to reality implies a contradiction. He then defends each of these premises, dividing the argument into two parts: one purporting to show that A-series distinctions are necessary for any series to be a real temporal series, the other purporting to show that the application of A-series distinctions to any series of real events implies a contradiction. Much of the subsequent discussion of McTaggart's problem attempts to criticize one of these parts and defend the other.

The argument for the claim that an A-series is necessary for time assumes that time unavoidably involves change. This does not necessarily mean that time is an aspect of change, but only that if nothing ever changed there would be no time. If time could consist only of a B-series, then change must be possible without an A-series. But, McTaggart argues, change requires an A-series. Therefore time cannot consist of a B-series alone; if it is real, time requires an A-series. Consequently, events are ordered with respect to a B-series only if they are located in a real A-series.

The reason a B-series alone cannot account for change is that if events or moments only stand in B-relations of earlier than and later than one another they stand in those relations permanently – nothing regarding those relations can differ in its characteristics. If two events or moments stand in a permanent relation, then standing in that relation cannot account for change. Change requires that some aspect of what changes differs with respect to at least one of its characteristics. As a result, neither B-relations alone nor a B-series alone can account for change. Moreover, because the only characteristics of an event that can change are its A-characteristics, an event can change only if there is a real A-series. For an event to undergo a real change some of its characteristics must change. Since McTaggart thinks of an event as the content of a temporal position, an event could not have different content and remain the same event. It would, then, be a mistake to say that it, that event, had changed if its content altered. Since an event is the content of a temporal position, only characteristics of that position could change. So all that remains that could change is whether the event is future, present, or past – its A-characteristics.

Alternative accounts of how change might occur in a B-series alone, McTaggart claims, must be rejected. Since positions in a B-series are permanent, change cannot consist in one event ceasing and another beginning. The first event retains its place in the B-series and has been replaced by, rather than changed into, the other event. Nor, for similar reasons, can change consist in an event in a B-series ceasing to be an event. Change, then, cannot consist in one event becoming another event, nor could occupying different moments in absolute time constitute change. Moreover, McTaggart replies to a claim made by Russell[4] that change consists in a proposition with different truth values applying to the same object, only differing in the fact that they apply at different (B-series) times. This supposes that change is a feature of the object that changes and not of events. McTaggart claims that this is not an adequate account of change, since the facts about the object do not change. If an object is P at time t_1, then that object is permanently P-at-t_1. There is no more change in a poker that is hot at t_1 and cold at t_2 than in a poker that is hot at one end and cold at the other. Both these facts are permanent features of the whole poker as extended in space or in solely B-series time. Genuine change requires temporal passage, i.e. requires that there be a real A-series.

Next, McTaggart argues that the application of A-series distinctions to reality implies a contradiction. Suppose that all events have real A-characteristics, i.e. that every event is either future, present, or past. Every future event eventually becomes present, and every present event eventually becomes past (assuming there is no last event). Consequently, every event possesses the A-characteristics future, present, and past. Future, present, and past, however, are mutually exclusive determinations, i.e. if an event has one of these it cannot have the others. Therefore the supposition that all events have real A-characteristics implies the contradiction that every event must possess all the A-characteristics – future, present, and past – and that since these characteristics are mutually exclusive no event can have them all. Hence, the application of A-series distinctions to reality implies a contradiction.

It may seem, at first, that this argument should not be taken seriously. The obvious reply is that no event has all of the A-characteristics at the same time; each event is first future, then present, then past in succession. But McTaggart thinks this reply produces a genuine paradox, i.e. it leads to either a vicious circle or a vicious infinite regress. The paradox arises because the reply must refer to a time series in its claim that no event has all of the A-characteristics at the same time. Either this time series is the same as the A-series the coherence of which the reply seeks to justify, or "at the same time" is a B-relation of simultaneity which assumes the existence of a B-series, or there is another time series – $time_2$ – from the perspective of which the reply claims that events in $time_1$ do not have all of the A-characteristics at the same $time_2$.

The first alternative is circular, since it assumes the coherence of the very A-series the reply seeks to justify. The second alternative only reproduces the paradox, since McTaggart has already argued that B-relations cannot constitute time without an A-series. This A-series is either the same as that which the reply seeks to justify or is a different A-series belonging to another time series, as in the third alternative. The third alternative, however, leads to a vicious infinite regress, since the second-order $time_2$ would also require an A-series. The reply would, then, be claiming that the distinctions of A-$series_1$ do not apply to events at the same A-$series_2$ time. But the argument that the application of the original A-series distinctions implies a contradiction now applies to A-$series_2$, reproducing the same paradoxical alternatives. To avoid circularity by appeal to yet a third time series requires yet another A-series subject to the same difficulties. This regress continues, McTaggart thinks, without ever reaching an A-series that is not contradictory.

Given the conclusion that any A-series is contradictory, why are we so strongly tempted to believe that the A-series distinctions can apply to reality? McTaggart claims that this temptation results from psychological features of experience. We have perceptions, memories, and anticipations, which differ qualitatively. These qualitative differences lead us to believe that perception has a feature, presentness, when we have the perception, which is replaced by pastness when we have a memory, and replaces futurity when we have an anticipation. We apply these features to events, calling present everything simultaneous with perception, and *mutatis mutandis* with past and future. This mistakes subjective features of experience for objective features of events. Moreover, we mistake qualitative differences in subjective experience for temporal features. Here again is a vicious circle: my perception is present when I have it means my perception is present when it is present. If we leave out the qualification "when I have it" the definition is false, since we each have many perceptions at different times. In that case every perception would be present, past, and future, and so contradictory.

Moreover, suppose that experience of time always grasps a temporal interval at once, what has been called the specious present.[5] Perceptions, then, may vary within the interval of the specious present. Consequently, for the present to be real it cannot be the same as the specious present. For in that case an event could be past when I experience it as present, or present when I experience it as past. Moreover, an event can be present in my specious present and past in someone else's. Were both perceptions accurate the event would be both past and present. This would be possible only if time, the A-series, were merely subjective. If we hold, instead, that the present is a point and not a duration, objective time, McTaggart argues, would still differ from perceived time. For, perceived time would have three durations – past, present, and future – while objective time would only have two: past and future separated by a point instant. In either case the duration of the objective present would be utterly unknown to us. McTaggart concludes that the unreality of time does not contradict our experience. Our experience of time would be mistaken whether time were objectively real or not.

Finally, McTaggart makes two ancillary points. First, he supposes that A-characteristics are relations rather than qualities. Yet he claims that were they qualities it would not affect the argument, since it would apply equally well to qualities.[6] Moreover, since the relations between events do not change, while the relations of the terms of the A-series do change, McTaggart thinks that one of the *relata* must stand outside the time series. Were this relation to constitute the A-series, then for the *relatum* that is outside of time to stand in relations of future, present, or past it must differ in some way with respect to the other *relata*. It would be difficult, he thinks, to find such an atemporal *relatum*.[7]

Second, the B-series, McTaggart argues, can be derived from the A-series and what he calls a C-series. In Hegelian fashion, he thinks that the series of inclusions of partial realities in a larger whole form a C-series. However, McTaggart's particular metaphysical predilections do not really matter. A C-series can be any permanent relations of existent things that do not by themselves constitute a temporal series, that can be generated by a transitive asymmetrical relation, and that can be misperceived as a B-series. If one and only one position of a C-series is present and that position passes along the C-series so that all the positions on one side have been present and all the positions on the other side will be present, this generates a B-series, i.e. a C-series becomes a B-series when combined with an A-series. An A-series adds change and a preferred direction to the C-series, which provides a permanent transitive asymmetrical relation. This is sufficient to transform the atemporal C-series into a temporal B-series, since a B-series is a temporal series, and so requires an A-series.

ANALYTIC ALTERNATIVES

Both because the argument is independent of his ontology and because he wrote in Britain, McTaggart's argument has been inherited by the analytic tradition that emerged in British philosophy at the beginning of the twentieth century. Analytic philosophy combines an empiricist stance in epistemology with a commitment to the detailed analysis of the meaning and structure of language as a tool in the resolution of philosophical problems. Time, even from McTaggart's perspective, bounds appearances. For McTaggart, this means that time is a misperception of a timeless reality. But for analytic philosophy, with its empiricist stance, McTaggart's position amounts to illicit metaphysical speculation. All intelligible discourse must somehow be tied to possible experience. Time is a boundary condition of appearances, and analytic philosophy attempts to understand time from within its appearances. Appearances consist in that which is given in perception and to which all meaningful discourse must somehow be referred. This empiricist approach refuses to accept the unreality of time. Time, as given in experience, is taken to be too obvious to deny.[8]

Comparison with Late Neoplatonism illustrates the difficulty. For Iamblichus, only the higher time was ultimately real. Its reality consisted in the articulations of the intelligible world laid out in sequential order from earlier to later, a B-series. The lower time attained reality only as the now point passed across and received reality from this real order.[9] For empiricists, a metaphysical B-series is anathema. So some sense must be made of how both A- and B-series can appear in experience. McTaggart's argument arises because of the difficulties of combining the dynamic with static aspects of our experience and our conception of time. Since time appears in experience it must be real. But if this is so something must be wrong with McTaggart's argument. Discussion among analytic philosophers attempts to explain what has gone wrong and how time, as it appears, ought to be conceived. An adequate explanation must account for both A-characteristics and B-relations. This has divided the majority of analytic accounts into two camps, A-theories holding that time must exhibit A-characteristics, and B-theories holding that only B-relations are real and that B-relations account for all A-characteristics.

For A-theories, temporal becoming is definitive of time; differences in A-characteristics inhere in all events. A-theories allege that this is indicated by significant ontological differences between past and future. Such theories accept some version of McTaggart's argument that the reality of an A-series is necessary for time. Hence, they also tend to accept that change requires the A-series and that the B-series alone cannot account for change. Moreover, they usually try to account for the B-series in terms of the A-series, often seeking to analyze B-relations as linguistic variants that actually describe or indicate A-characteristics.[10] Consequently, they reject the arguments that the application of A-series distinctions to reality implies a contradiction. B-theories accept these arguments. Instead, they hold that B-relations are genuine temporal relations and that objective time consists solely of the B-series. B-theories maintain that change can be accounted for by a B-series alone. They treat temporal becoming as a subjective result of B-relations between events and their perception. They tend to eliminate objective A-characteristics by analyzing them in terms of B-relations.

To gain further insight into the difficulties raised by McTaggart's problem, in the rest of this chapter we shall examine issues concerning change and the transient aspects of time, and then we shall consider the status of McTaggart's paradox and whether the existence of an A-series must be contradictory. Chapter X will take up issues of representation, existence, and temporal experience.

TRANSIENCE AND CHANGE

While we undoubtedly observe change, it is a matter of contention how change is related to time and whether transience, temporal passage, is a feature of time or of what changes. A-theories adhere to the transience of time. They hold that no analysis of time can ignore the evident experience that time passes. With the passage of time one phase of experience continuously supersedes another. Passage is not a feature of the contents of experience. Instead, A-theory claims, passage is essential to distinguish temporal succession from a temporal interval. A-theory conceives of temporal succession as the change of the present and what is present into the past. Things, events, and time intervals become older regardless of whether they undergo qualitative changes.[11] It is in this sense that McTaggart writes of events changing. Events recede into the past but do not change qualitatively; what has happened has happened. Becoming older is pure transience, not a change in qualities; when it obtains, each moment or time interval becomes present at some time and then recedes into the past. A-theories connect temporal transience closely with the presence of experience. Pure transience obtains whenever change is experienced. Experience is always present; whatever is given directly to consciousness occurs in the present.[12] A-theories hold that this connection with consciousness does not necessarily impugn the objectivity of A-characteristics. A clock can announce the present time without any conscious experience.[13] A-theory takes the presence of experience as supporting its claim that the present is a real aspect of time.[14] If the B-series cannot account for the experience of temporal transience it fails to explain a key feature of time.

Despite its experiential immediacy, temporal transience is difficult to describe. The metaphor of the flow of time suggests that pure transience is a kind of motion. Both A- and B-theorists often picture the present as the now point sliding along the sequence of events. Again, this evokes the diagram of the Pseudo-Archytas, where the moving now advances along the B-series, shifting its position from earlier to later. Without recourse to the metaphysics of a higher reality it is difficult to explicate how the conditions of motion can apply to the moving present. It is difficult to conceive of motion that does not presuppose time. Motion requires a contrast between that which moves and a background time series against which it moves. However, it is not clear how there could be a background time series for the now to move against without producing a variant of McTaggart's paradox. If the background is a B-series, then a B-series could exist without an A-series. If, instead, every time series must include an A-series, the moving now will demand either a circular appeal to itself to explain its motion or an appeal to yet another times series, which must itself include an A-series, thereby producing an infinite regress of time-series. It seems that nothing can move solely in time.

That motion requires contrast also suggests that motion can occur at different rates, and so presupposes a time series. If the now could move along the sequence of events it would make sense to ask how fast it moved. Yet that question seems senseless because the rate of motion is a ratio between some sequence, e.g. of places, and the time it takes for that sequence to elapse. But if that sequence is itself a time series, then the question asks how great a lapse of time does the now traverse in a unit of time. Presuming there is only one time series, this again generates a vicious circle or, if there is a second-order time in which the now of the first-order time moves, an infinite regress. This argument gets support from both B-theorists and some A-theorists.[15] They agree that it is an error to treat becoming present as a change in quality like becoming hot. The argument shows that pure transience cannot be a form of qualitative change. Instead, qualitative change presupposes pure transience. Motion is just one form of qualitative change and pure transience cannot be explained in terms of it.

B-theory sees this argument as part of a refutation of the reality of an A-series. A-theory may respond by accepting the argument while denying that pure transience is properly described in terms of motion and the moving now. Consequently, no rate can be assigned to temporal transience. The problem remains, then, of how to understand what it means for something to become present. Some A-theorists argue that the B-series treats the whole of time as coexisting, or at least as rendering all moments equal, and thus failing to do justice to temporal transience, which requires that the special status of the present moment be explained. More needs to be said than that to become present is to happen or come to pass.[16]

Alternatively, some A-theorists try to make sense of the rate at which the now moves. If a rate of change is a ratio between anything at all and a time interval, then there can be a ratio between a time and itself. Whatever grows older does so at a rate of one time-unit per time-unit; this is no more mysterious than my being my own sibling, since I have the same parents as myself.[17] But such an answer seems an unilluminating tautology. It is trivially true, and compatible with a B-theory, that a time is privileged at that time. It may be possible to assign a contingent rate to the motion of the now if we can distinguish contrasting temporal-series. If there were two time series the now in each series might move reciprocally with respect to the other. The rate of flow would be the ratio of the time elapsed in one series to the time elapsed in the other. Were this scenario even possible it would show that the moving now is at least a coherent option. However, it seems ad hoc merely to posit both an ordinary time and a single meta-time. It is difficult to differentiate contrasting time series without entering into complex disputes about other matters, particularly issues concerning existence.[18]

B-theorists question the motives for endorsing the moving now, attributing it to our psychological limitations and the hypostatiza-tion of features of our psychological states.[19] As our perceptions change we change our beliefs to reflect differences in our temporal relations to the objects we perceive. So, for instance, on the second day of January we can no longer believe truly that today is New Year's Day. We change our tensed beliefs to accommodate changes in their truth values. These are not changes in the objects of our beliefs, but they explain the psychological states hypostatized as the moving now.[20]

The moving now also seems to require that events can happen to events. For events would acquire presence when the now moved across them, and lose it when it had passed. The acquisition and loss of presentness would comprise second-order events. However, rejecting this conjecture may dispose of McTaggart's paradox on the supposition that the infinite regress can get started only if we try to treat the acquisition and loss of presentness as a kind of event.[21] In attempting to explicate change, B-theory argues against the coherence of events happening to events. Many differences are due to difference in the relation in which something stands to something else. I may become closer to you when you walk toward me. This is not a change in me in that it is not the result of any causal forces that go through me. Likewise, B-theory maintains, events may stand in different relations to other events, but do not really change.[22] B-theory holds that there is no special moment that is present, nor any recession into the past that could constitute a real change in either times or their contents. What is correct in the notion that time passes is that events occur in sequence.[23] That time forms a B-series, then, is sufficient to account for the temporal aspect of all real changes. Things change: flowers bloom, metal rusts. A change in a thing consists in its possessing incompatible properties at different B-series times. This explication of change presupposes a clear distinction between things and events.

Things retain their identities over time, while changes are events that happen to things. If no thing retains its identity when an event occurs, the event does not count as a change.[24] A thing comes to be insofar as it exists at some time when it did not exist at an earlier time. Coming to be and ceasing to be, then, are events but not changes, since nothing in these events retains its identity while successively acquiring different properties. As identities preserved through difference, things exist as wholes at any given time; it is the whole thing that endures over time. As a unity, a thing forms a nexus of causality; any real change in a thing's properties arises from its compliance with causal laws. A thing's properties become manifest in such changes. Division of a thing into parts would be arbitrary if they were not causally identifiable independently of the wholes they comprise. A thing's parts must be together at one time for them to interact causally and produce changes in the whole thing. Consequently, things have no temporal parts.[25] Things, then, can undergo real changes, i.e. changes that have effects attributable to the causal properties of that thing. Merely becoming older, on this account, is not a real property of anything since, even according to A-theory, it has no qualitative effects. B-theory will claim that becoming older simply consists in some facts about B-series distances. That I am older than I was last year consists in there being a greater distance between this year and my birth than last year and my birth.[26]

Every change is an event, but not every event is a change in some thing; a flash of lighting, for instance, is a single event but not a change in any one thing.[27] Only things can change or become different. B-theory holds that pure becoming makes no sense; no objective A-series exists and so future events cannot become present and then past. For a thing to become, it must differ qualitatively due to the intervention of an independent cause. This is why things are not just sequences of events; they retain their properties unless affected by some intervening cause. Things change because they retain their identities through qualitative change, but events cannot undergo change. For events to change, events would have to be able to happen to events. This would require that an event retain its identity over time. The ordinary use of "event" in English includes both instantaneous happenings and those that endure: limits and processes.[28] Starting, arriving, and winning, for instance, realize limits; they happen at a time, but take no time to happen. Accelerating, moving, and fighting exemplify processes; they happen over time intervals and can be analyzed into phases. However, neither limits nor processes retain their identity over time. Although a limit retains its identity and may be re-identified at a later time, its identity does not extend over any time interval. Limits, then, cannot count as undergoing real changes, for they cannot have different properties at different times. And, while a process extends over a time interval, there is no one time at which it exists as a whole. Processes, unlike things, have temporal parts. So processes do not change; they merely differ in their temporal parts.[29]

B-theory, then, holds that change is a feature of things in time. It rejects McTaggart's argument that the B-series cannot account for change because it contains only unchanging relations and permanent facts about things. The permanent possession of a property does not follow from the timeless truth of the proposition that something possesses that property. While it is timelessly true that a poker is hot-at-t, it does not follow that the poker is permanently hot-at-t. McTaggart's reply to Russell misses the point: it is things which change, not facts. The B-series orders changing things according to relations of earlier, later, and at the same time. It is no more subject to change than are numbers. Moreover, a poker hot at one end and cold at the other differs spatially from one end to the other. Insofar as an independent cause produces this difference, the difference is a real spatial change, just as the change from hot-at-t_1 to cold-at-t_2 is a real temporal change.[30]

THE STATUS OF THE PARADOX

In addition to making sense of temporal transience, A-theory must come to terms with the paradox McTaggart alleges arises in the conception of an A-series. McTaggart's argument claims that every event is past, present, and future, and yet these determinations are mutually exclusive. Both these claims cannot be true. The reply that these determinations do not apply to events at the same time seems to reintroduce a time series, producing a vicious circle or a vicious infinite regress of times. A-theory must explain why this argument fails to reduce A-series distinctions to incoherence. To do this, A-theories have pursued two strategies: some argue that no contradiction, and so no regress, really arises, while others argue that the regress does arise but is not vicious. To explain and avert the fallacy it finds in McTaggart's argument, A-theory turns to the analysis of temporal language. Although they differ regarding where the error lies, many A-theorists suppose that the error arises because of some systematically misleading use of language.

The argument that no contradiction arises attempts to distinguish the conditions in which A-characteristics are compatible from those in which they are incompatible. A-characteristics are incompatible when applied to events or moments timelessly or simultaneously, but are compatible when applied successively.[31] C. D. Broad (1887–1971), who originated this and many of the now standard lines of argument regarding McTaggart's problem, attributed the fallacy to mistaking transience for a form of qualitative change. What Broad thought had misled McTaggart was his attempt completely to replace tensed copulas ("is," "was," "will be," etc.) with tenseless copulas and temporal adjectives ("is present," "is past," "is future," etc.).[32] To say that an event is (timelessly or simultaneously) both past and future is contradictory. This contradiction arises only if the adjectives "past," "present," and "future" apply to moments or events timelessly or simultaneously. Such adjectives, Broad claims, can only legitimately apply successively. The regress, therefore, results because the use of tenseless copulas requires timeless or simultaneous application of temporal predicates to perform the functions of tense. Incompatible applications of temporal predicates can be avoided only by reintroducing a tensed copula, which, when eliminated, will in turn reproduce the difficulty and generate a regress. Accordingly, Broad argues, the regress need not get started at all. We need only reject the project of eliminating tensed copulas and accept that tense cannot be eliminated.

Later A-theorists extend this claim by pointing out additional features of the linguistic expression of A-characteristics. Prepositional phrases ("at a past time," "in the future," etc.) in ordinary English generally add emphasis rather than replace tensed verbs. Moreover, if instead of using adjectival phrases one used adverbial phrases to modify verbs to make new statements out of other statements, then such phrases can be added without producing a vicious regress. Hence, "It is two years since E" can be modified to become "It is two years since it was two years since E," producing a new claim and no regress. Adverbial phrases like "it was (is now, will be) the case that . . . " function like "it is not the case that . . . " They modify whole statements, by modifying their verbs, and are capable of expressing any required degree of temporally complex facts. No regress results, since temporal adverbs need not be used in different senses when iterated, and no additional time series need be supposed.[33] This avoids confusing constructions like "E is (tenselessly) present in the future." Instead, we may say that it is now true that "it will be the case that E occurs" and "it is now the case that E" will be true at some future time.[34]

The adverbial analysis allows the elimination of tensed verbs. If tense can be accounted for by adverbial phrases, verbs themselves need not have tenses.[35] Yet, while tense need not be a feature of verbs, transferring tense to adverbs does not eliminate it. However, recourse to adverbs does avoid treating A-characteristics as properties of events or moments. Unfortunately, if the adverbial analysis suggests that A-characteristics are properties of facts it makes little progress. A-characteristics cannot be both properties of something (things, events, propositions, facts, etc.) and incompatible.[36] First, no instantaneous event or moment can have incompatible properties. Second, temporally extended particulars can only have incompatible properties at different times. The former makes it impossible for events and moments to change their A-characteristics, e.g. for events merely to become older. The latter treats past, present, and future not as properties, but as relations between particulars and times. This reintroduces B-series times. For were particulars related to A-series times these moments would have to change with respect to their A-characteristics, and a vicious circle or infinite regress would follow.[37]

The second strategy for avoiding McTaggart's paradox accepts the infinite regress while denying that the regress is vicious. Entirely vicious infinite regresses reproduce the difficulty that generates them at every stage. Entirely benign infinite regresses contain no such difficulty at any stage, and a harmless relation generates their infinite progression.[38] On one interpretation of McTaggart's regress, a tenseless description generates a contradiction at each stage that is then resolved by a consistent tensed description. The regress is, then, peculiar in that it alternates between inconsistent and consistent claims. There would only be inconsistencies at alternate stages, and it is not immediately obvious why it should stop at any given stage or why the regress should continue past the first consistent stage.[39] Broad, for instance, argued that the original description is consistent, so no regress gets started. The regress seems to arise from producing a contradiction by illicitly requiring a tenseless description.

This interpretation attributes the paradox to a preference for tenseless descriptions. This linguistic point, however, does not address the substantive issue about time. The paradox is a consequence of the infinity of A-series. There must be a different A-series at each moment. For, A-characteristics of events or moments change, e.g. a present event becomes increasingly past, and which moment is present continuously changes. Hence, each A-series must be relativized to some time. If that time is itself identified by its A-characteristics, then those characteristics must also be relativized to some time. Either these relativizations regress infinitely or there must be an absolute A-series.[40] If the latter, then there would be a series that can be tenselessly or simultaneously described. This description would then be inconsistent, since every item in the series would be past, present, and future. If the relativizations are allowed to regress infinitely, then, for an A-series to exist, that regress would have to be benign.

Early attempts at accepting the regress mistook the limit of an infinite series for an end-point.[41] Instead, it may be possible to generate the regress without generating any inconsistencies. There is nothing, in itself, objectionable about an existent infinity of A-series if it contains no contradictions. Attempts to deny the contradiction seem forced to reject the view that A-characteristics, to retain their incompatibility, are properties of anything. However, if A-characteristics can be both properties of events and properties of properties, then it may be possible for them to be incompatible and not lead to contradiction. If pastness, presentness, and futurity can inhere in the series of events successively, no contradiction arises. To the question of when these characteristics inhere in events, the answer is that they inhere at present. This presentness is a second-order property; it inheres in the first-order properties pastness, presentness, and futurity. This does lead to an infinite regress of properties of properties, but does not posit more than one time series as the bearer of infinitely complex properties. Such a regress is not, prima facie, vicious insofar as the inherence of properties in properties is not by itself objectionable. It is possible to describe any given level of the regress, and, so described, each level of properties entails and analyzes the level below. Hence, "presentness now inheres in E" entails "presentness inheres in E" but not "E was (will be) present." For each attribution of an A-characteristic an infinite set of entailments follows, *mutatis mutandis*. Each iteration uses tensed verbs, and adverbial modifications indicate the inherence of some property. No relations of contradiction and resolution seem to arise to generate a vicious regress.[42]

This infinite inherence of properties in properties does, however, privilege presentness. That privilege again threatens the coherence of an A-series. For an A-characterization of any given complexity eventually inheres in presentness. For instance, that E will be past entails that presentness inheres in the inherence of futurity in the inherence of pastness inhering in E.[43] This means that the second-order inherence of futurity in the pastness of E is now present. But then the inherence of this pastness must be both present and future, since it is now present. Again, a contradiction arises and the regress appears vicious.

These considerations are still matters of contention. At best, should an infinite series of inherence of A-series properties prove consistent, it would show that an A-series is possible. To do this it would have to demonstrate under what conditions descriptions of A-characteristics could be true. There would have to be an infinite set of such conditions. It would be a further task to show that these conditions actually obtain.

CHAPTER X

Tense and existence

The Moving Finger writes and, having writ,
Moves on: nor all your Piety nor Wit
Shall lure it back to cancel half a Line
Nor all your Tears wash out a Word of it.

Omar Khayyam

Analytic philosophers attempt to give an empirical account of both the changing present and the direction of time from earlier to later. Experience manifests both the A- and B-series, yet McTaggart's argument raised doubts about how these series could exist. Analysts debate these ontological matters by way of the relationship of time to its linguistic representation. (We have already seen the purported error in McTaggart's argument attributed to misleading uses of verbs.) More positively, analysts look to linguistic devices such as tense, and the analysis of the sense and reference of A- and B-series descriptions to test the coherence, correctness, and relations between these representations. We shall now examine some of this analysis of temporal language, some issues regarding the relation of time to existence, and then consider the experience of time.

TENSES AND TRUTH VALUES

Among analysts the relationship between language and the world is itself the subject of complex debates. Regarding the connection of time with language, A-theory tends to appeal ontologically to those impressed by everyday experience and a commonsense world-view, and linguistically to those impressed by natural languages, while B-theory tends to appeal ontologically to those who privilege natural science and a scientific world-view, and linguistically to those attracted to formal languages. Some theorists diverge from these tendencies, but such tendencies have their roots in theoretical considerations. Natural languages, at least Indo-European languages, represent time via tense. Tensed sentences represent time as an A-series, and in their use they ascribe A-characteristics successively and hence compatibly. So A-theorists often argue that since some tensed sentences are true, time, as ordinarily perceived, possesses A-characteristics.[1] B-theorists, conversely, tend to argue that formal languages better express the results and methods of scientific theory. Although formal languages may differ from natural languages, they are able to remove ambiguities and fix truth values timelessly. In this way, formal languages accord with physics, which excludes A-characteristics.[2]

Consider, again, the problem of eliminating tensed copulas in favor of tenseless copulas and temporal adjectives. A-theorists aver that tense cannot be eliminated and that attempts at elimination violate rules of linguistic usage, that the ordinary linguistic rules governing tensed sentences require that true descriptions of events must be tensed. To ask "When is E present?," unlike "When was E present?," would be redundant unless the copula were tenseless.[3] So, while every event either is, was, or will be truly describable as present, this gets expressed through tensed verbs rather than temporal predicates. The copula in "E is present" is tensed, since the same sentence when uttered at different times does not express the same fact. At a later date the fact that E is now present will be expressed by "E was present." The tensed sentences "E is present" and "E was future" are not contradictory; if the event is present both are true.[4] Moreover, ordinary usage requires that different tenses apply successively.[5] This grounded the A-theorists' reply to McTaggart. The use of tenseless verbs and temporal adjectives tends to make successive events seem contemporary. A-theory protests that such attempts to eliminate tensed sentences in favor of timeless descriptions unnecessarily produce paradox. Ordinary linguistic usage already has a coherent logic, since tensed sentences contain a consistent set of tautologies and rules of use.[6] It is never jointly true to say that E was, is, and will be present, only that it was true that E is present, or that it is now true that E is present, or that it will be true that E is present.[7]

This last claim presumes that descriptions may vary in truth value when uttered at different times. This feature of ordinary tensed language can be used to distinguish tensed from tenseless sentences.[8] A sentence is tensed if it makes a claim that differs in truth value at different times, while it is tenseless if it retains the same truth value at any time. Tenseless sentences are, in this sense, freely repeatable. Tensed sentences are not freely repeatable, since they have different truth values at different times.[9] On this view, the quarrel whether time is properly represented by tensed verbs or by temporal adjectives operates at the wrong level of analysis. Tense applies to sentences rather than to their grammatical components. The adverbial analysis that treats tenses as sentential operators already points in this direction.[10]

In opposition to this view, many philosophers hold that truth must be timeless, sentences can never change their truth value over time, they cannot become true or become false. Standard formal logic accepts this and fixes the truth values of sentences. This means that every sentence that seems to change truth value over time must be reconstrued with an added time index. So, for example, "The cat is out" gets read as "The cat is out at 11.00 p.m." Addition of a time index provides a rule-guided and uniform way of taking temporal differences into account. Times become the values of variables of quantification.[11] Linguistically, this approach tries to eliminate tense in favor of temporal predicates. Alternatively, a time index can act as a global constraint on the assignment of meaning to sentences.[12] In either case, truth is timeless and the question of when a sentence is true is senseless. While this is at odds with ordinary ways of speaking, it accords with the context-free descriptions favored by natural scientific theory. Statements of the laws of physics, for instance, hold true regardless of particular circumstances.

TRANSLATION AND REDUCTION

Analysts transform the question of whether B-relations are sufficient for time or whether A-characteristics are also necessary into the question of whether tensed sentences can be eliminated without residue in favor of tenseless sentences. Initially, B-theory attempts to eliminate tense and to construct systematic translations of tensed into tenseless sentences. A-theory not only resists such translations, but also attempts to translate expressions of B-relations into the language of A-characteristics. If the language of one type of series can be translated without residue into the other, then only the translating language is necessary for describing time. This eliminates the need for the translated language, suggesting that the series described by that language derives from the series described by the translating language. If tensed language can be eliminated and everything that can be said in tensed sentences can be said without them, then it is thought that tense does not ascribe real characteristics to time and, consequently, that A-characteristics can be reduced to B-relations. If expressions of B-relations can be translated without residue into sentences using only the language of A-characteristics, then it is thought, with McTaggart,[13] that B-relations can hold only if A-characteristics already apply.

An adequate reduction must account for the distinguishing features of the reduced series. Hence a reduction of B-relations to A-characteristics, for instance, must account objectively for the structural order of temporal distance and direction. A translation of sentences expressing B-relations into tensed sentences, then, must unambiguously express an unique B-series, even though the tense, and hence the truth values of the translating sentences, changes continuously over time. Since a new A-series arises at each moment, an adequate description of A-characteristics must contain a variable temporal index indicating which moment is present. A translation of tensed sentences into tenseless ones must analyze changing tenses exhaustively and unambiguously into sentences that retain their truth values over time. Descriptions expressing only B-relations must not change truth values or contain a temporal index. A reduction of A-characteristics to B-relations, then, must account objectively for the appearance of temporal passage.

A-theory claims, instead, that expressions of B-relations can be translated without loss of meaning into tensed language. Richard Gale, for instance, translates "E_1 is earlier than E_2" as E_1 is past and E_2 is present, or E_1 is past and E_2 is future, or E_1 is present and E_2 is future, or E_1 is more past than E_2, or E_2 is more future than E_1.[14] Similar constructions can be adduced as translations of "E_1 is later than E_2" and "E_1 is simultaneous with E_2." The disjuncts employing degrees of pastness and futurity are necessary to account, respectively, for cases where E_1 and E_2 are both past or both future. The analysis claims to reduce B-relations to A-characteristics, by making it possible to describe the generating relation of the B-series (. . . earlier than . . . , or . . . later than . . .) entirely in A-series terms. Yet Gale acknowledges that events may admit of relative degrees of pastness or futurity. This seems to require that the A-series also be generated by B-relations.[15] The claim that E_1 is more past than E_2 is tensed, since it relies on a temporal indexical expression and so can have different truth values at different times.

However, it is possible to analyze this claim as a conjunction of "E_1 is earlier than E_2" and "E_2 is now past." That the first conjunct expresses a B-relation threatens the translation with circularity. The threat is not vitiated by the fact that conjunctions of A- and B-statements are always tensed. Gale accepts that "E_1 is more past than E_2" entails "E_1 is earlier than E_2" and "E_2 is now past," but denies, without argument, that this entailment is an analysis.[16] Consequently, even were the translation adequate it seems difficult to avoid accepting the reality of B-relations.

Furthermore, while each of the disjuncts in the translation of "E_1 is earlier than E_2" is tensed, the whole disjunction is not tensed; it retains its truth value at all times. Of course, any analysis must be at least logically equivalent to the claim analyzed, so if the latter is true at all times the former must be also. But in that case the equivalence obscures whether the translation effects a reduction of B-relations to A-characteristics. Partly, the difficulty concerns the timelessness of the B-series. B-theory maintains that the B-series is a temporal series in its own right, independent of the A-series. The B-series orders events according to transitive asymmetric B-relations, providing events with an intrinsic temporal direction. This order does not change with time, so true descriptions of it retain their truth values at every moment; the facts about temporal order are not temporally ordered.[17] Consequently, descriptions of B-relations need not be tensed.

A-theory objects that these timeless descriptions are insufficient to account for temporal passage. Each A-series indexes one moment as present and determines other moments as past or future relative to that moment. The present appears to highlight one moment of time after another. This presence of appearances inclines A-theory to treat the present as one of three time determinations, where B-theory recognizes only two.[18] Although the present moment somehow accompanies the presence of experience, A-theory holds that which moment is present is neither arbitrary nor subjective. B-theory, conversely, seems forced to treat the present either as applying equally to every moment or as subjective. Descriptions using only B-relations cannot pick out a unique moment as present.[19] The B-series treats all moments as equivalent, differing only regarding their position in the series. From these B-series coordinates it is impossible to determine whether an event occupying them is present.[20] B-theory tends, therefore, to relegate presence to subjectivity.[21]

Initially, B-theory responds to these difficulties by attempting translations of representations of A-characteristics into representations of B-relations. Such translations aspire to eliminate the need to describe time in terms of A-characteristics, and especially to eliminate the need to favor some moment as present. B-theory resists the tendency to treat A-characteristics as properties of events or moments. Instead, it tries to explain the comprehensibility of tensed language while ascribing only B-relations to events and moments. Eliminative translations, through either a date analysis or a token-reflexive analysis, would reduce A-characteristics to B-relations. If such translations could be pulled off, B-relations would be sufficient to constitute time. If no representations of A-characteristics are necessary and all facts about objective time can be described entirely in terms of B-relations, those facts need not include A-characteristics.

The date analysis proposes treating B-series positions, dates, as values of variables of quantification.[22] Tensed sentences are then eliminated by replacing them with tenseless sentences that retain their truth values because they explicitly mention the dates to which they apply. So, for example, "A storm is now occurring" gets translated as "A storm occurs at 2.00 p.m. on July 1, 1998." This translation requires only the B-relation that the storm occurs at a time some specified quantity of time later than the event that functions as the origin of the calendar. Similarly, a date analysis translates "*E* is past" into "*E* occurs before 2.00 p.m. on July 1, 1998," where the verb is tenseless. The translation purports to preserve the objective content of A-statements without attributing A-characteristics to objective time.[23] It treats dates as attributes of events or things that stand in B-relations to one another, and B-relations as the only objective aspects of time.

Such an analysis treats things and events timelessly; it regards a storm on July 1 as a sort of storm. However, that a storm occurs on a specific date answers the question "When did the storm occur?," not the question "What kind of storm was it?"[24] Defenders of the date analysis recognize this but find no difficulty in it. They appeal to the representations of objects in the space–time of contemporary physics. Physics provides a factual description of the natural world that need not bend to the vagaries of ordinary language.[25] This description requires only B-relations to locate things and events in time, and so supports the claim that time objectively includes only B-relations. More problematic is the difficulty that a date analysis gives no motivation for selecting one date rather than another to relate to the date of an event when translating representations of A-characteristics. If "*E* is past" means that *E* occurs before, for example, 2.00 p.m. on July 1, 1998, the analysis gives no reason to select that date rather than another. A-theory attributes this to neglect of objective A-characteristics. For many B-theorists it recommends a token-reflexive analysis.

Token sentences are particular things or events, e.g. inscriptions or utterances, that instantiate a general type sentence.[26] One type sentence may be produced on different occasions but be instantiated by a different token sentence each time. Token sentences, then, are particulars with definite dates. The token-reflexive analysis selects as the present, relative to the content of the sentence, the date its token is produced. So, a token-reflexive analysis translates "*E* is present" as "*E* occurs simultaneously with this token sentence," "*E* is past" as "*E* occurs before the utterance of this token sentence," "*E* is future" as "*E* occurs after the utterance of this token sentence," and similarly for other tensed sentences. In each case the translating verb is tenseless and the demonstrative points to the token sentence. The analysis accounts for the use of different tenses by appealing to differences in the B-series positions of token sentences and their B-relations with their subject matter.[27] A reflexive reference to the token sentence combined with an expression of the B-relations between the date of that token sentence and the date of the event about which it speaks replaces tense. For instance, a time interval may recede into the past in the sense that the B-series distances between that time interval and progressively latter token sentences increase. Hence this analysis purports to eliminate tensed sentences in favor of tenseless sentences.[28]

A-theory objects that any B-theory description, including the token-reflexive analysis, regards all moments as equal. A-theory views this as failing to account for the special status of the present. For the token-reflexive reference to succeed, the token sentence must presuppose its own presence, since it can only make reference to itself while it occurs.[29] B-theory defends the equivalence of moments, pointing out that each moment will be equally present relative to some simultaneous token sentence. A-theory claims that *E* is simultaneous with this token sentence should, but does not, entail that the token sentence is present.[30] B-theory responds that this begs the question since the present including the presence of the token sentence is not an objective feature of time. That the entailment of objective A-characteristics fails marks, instead, an advantage of the analysis.[31] The B-relations between events and sentence tokens are all there is to the objective present. Nor does the token-reflexive analysis require that any tokens actually be produced. If no token sentence occurs, that does not mean that nothing occurs in time, just that nothing could stand in a B-relation to a non-existent token.

Despite these considerations, there are compelling reasons why tensed discourse cannot be eliminated. Tensed sentences do not, prima facie, judge about themselves, nor do they seem to use tense-less verbs. While they may not assert that their content is simultaneous with their utterance, tensed sentences do have a use in natural languages, and that simultaneity functions as a rule for their use.[32] Instead of referring to the time of the utterance, the relation between events and token sentences functions as a circumstance determining the sentence's tense.[33] This circumstance constrains but cannot eliminate the indexical element in tense. No combination of sentences without an indexical element can entail a sentence containing one.[34] For example, the sentence "It is now 1998" does not mean the same as "The sentence 'It is now 1998' occurs in 1998", since the latter is true at any time while the former is true in 1998, but not otherwise. Like written sentence tokens and beliefs, the truth values of tensed sentence types vary with their temporal position.[35]

The indexical element of tensed language cannot be eliminated by token-reflexive analysis. For when one says "Thank goodness that's over!" it is absurd to suppose that one is thanking goodness that the date of the event's conclusion is (tenselessly) before the date of the utterance.[36] We use tensed language for practical purposes, to communicate systematically related temporal perspectives and to coordinate our actions. Tensed sentences often function to tell someone what time it is. Tenseless sentences, like "*E* occurs in 1998," cannot do this. Consequently, tensed sentences must be used to coordinate actions. The indexical element in tensed communication exerts the pragmatic force of looking at one's watch without requiring us to take the time to do so. Participants in cooperative action must share the same temporal perspective or be able systematically to coordinate their perspectives. So, they must use sentences whose truth values are sensitive to these perspectival differences. To perform any action successfully at a particular time requires that the actor have a true belief that will become false after the appropriate time. Such beliefs can be expressed and communicated only with tensed language. Coordinated and successful action, then, requires the pragmatic force of tensed language.[37] These uses prevent the elimination of tensed sentences regardless of other aspects of their meaning that may be captured by the token-reflexive analysis.

That tensed language has an ineliminable pragmatic role does not thwart B-theory. The meaning of ordinary temporal discourse arises in the context of concrete situations of use. In specific everyday practices we never need to assert that something is past and future at once. Conversely, the abolition of tenses suits natural scientific purposes. Natural scientific language ignores local circumstances, seeking greater generality and objectivity. Some thinkers claim that A-series descriptions and B-series descriptions can be translated into one another, but that their context of use constrains each sort of language. Both are limited by their context of use. Confusion and contradiction arise only when the two sorts of descriptions are mixed. A-series descriptions are appropriate in ordinary practice. B-series descriptions belong in natural science.[38] Yet if natural science discovers the real structure of the natural world this mutual-translation account favors B-theory, which asserts that real (natural) time is B-series time.

Recently, however, B-theorists have conceded that no adequate translation of A-series language into B-series language is possible. Translation fails not only for the reasons already suggested, but also because tenseless language cannot always be substituted into statements of propositional attitudes without altering their truth values. Someone can believe that an event is happening now without believing that it is happening at B-series time t_n. Moreover, one can know a priori that t_n is t_n, but not that t_n is now.

This new B-theory denies that translation between the two languages is necessary to determine the real nature of time. The new B-theory rejects the notion that ontological conclusions follow from the analysis of meanings, that the structure of language must share its form with reality. Instead, it points to recent theories of reference that institute a break between linguistic form and ontology. Such theories deny that those conceptual aspects of meaning which raise problems of translation determine that to which linguistic expressions refer.[39] This separation of questions concerning meaning from questions concerning the objects of reference bypasses the debate about translation. Translation of tensed into tenseless language is not necessary for the truth of B-theory. What is necessary is that there be an objective B-series, and that both tensed and tenseless language refer only to B-relations. Tensed language is necessary for practical purposes, but those purposes can be fulfilled even though tensed sentences really only refer to objective B-relations. Differences in tense do not denote ontological differences.[40]

While tensed language may be made consistent by certain rules for its use, the new B-theory claims it cannot be consistently true of time in nature. For A-characteristics to obtain in nature they would have to make tensed sentence types true or false. So, for example, any sentence token "*E* is future" will be true if and only if *E* is future. But tensed sentence tokens must, by definition, vary in their truth value at different times. Even if it is now true that *E* is past, some tokens of "*E* is past" may be false. A tensed sentence type will not retain its truth value over time, and so cannot be made true by an A-characteristic obtaining in nature. Were a sentence type made true thus, it could not be tensed. A B-series therefore comprises natural time.[41]

The new B-theory explains all objective truths about time, including its apparent A-characteristics, by appealing to the objective B-relations that make statements about time, including tensed ones, true or false. The truth or falsity of tensed sentences may be explained not by a token-reflexive account of its meaning, but by the objective B-relations that hold between the B-series positions of events and sentence tokens. So, for example, "*E* happened last year" will be true if and only if *E* occurred in the year before the year in which the sentence token was uttered. Hence the truth values of various token sentences will change depending on the objective differences between the B-series positions of the tokens and events. This accounts for the defining characteristic of tensed sentences – that they may vary in truth value when uttered at different times. Yet it does so without reducing tensed to tenseless sentences. Instead, it shows how objective B-relations make sentences true or false, and why no objective A-characteristics need be assumed to account for their truth values. Objective B-relations make tensed sentence tokens true or false, and differences in the B-series positions of these tokens account for differences in their truth values. Objective time consists in B-relations regardless of how it is represented.[42]

EXCURSUS: SPACE–TIME: THE THEORY OF RELATIVITY

The most important twentieth-century development in the physics of time is the theory of special relativity proposed in 1905 by Albert Einstein (1879–1955). The theory reconciles two seemingly inconsistent postulates:

1 The laws governing change in physical systems will be the same in any coordinate system in uniform translatory motion.
2 In a vacuum light travels at a constant velocity, c, regardless of the state of motion of its source.

The second postulate generalizes experimental results achieved in 1887 by Michaelson and Morley showing that the velocity of light measures the same both in the direction of the Earth's motion and perpendicular to it. This led Einstein to reconceive basic temporal relations in terms of the connection, by light rays, of two spatially separated clocks. Suppose a light ray traveling from clock A to clock B is reflected back again. If it begins at a time t_A measured by clock A, arrives at time t_B measured by clock B, and returns to A again at time t_A' measured by clock A, then if $t_B - t_A = t_A' - t_B$ the two clocks are synchronous (Figure 5). The synchronization relation is reflexive, transitive, and applies to any number of spatial points. This definition assumes that the speed of light is the same in both directions and is therefore a universal constant, c, equaling twice the distance from A to B divided by the difference between the times as measured by clock A.[43] Whether events are earlier, later, or simultaneous is a function of the relative states of motion of their sources, and lacks any absolute determination.

The first postulate, called the principle of relativity, concerns causal relations between events. It assumes that the physical world forms a unified system connected by causal laws, and asserts that these laws are invariant between spatio-temporal coordinate systems. This would seem to conflict with the second postulate. For at low velocities measurements of velocity appear to be governed by principles of simple vector addition, and laws governing measuring devices do not vary with differences in uniform velocity. It seems, then, either that measurements of the speed of light should vary with the motion of the measuring instruments in different coordinate systems, or that the laws governing measuring devices would have to differ in different coordinate systems.

The theory of relativity removes the apparent discrepancy. It accepts the two postulates, but abandons the simple vector addition of velocities. The theory explains why the speed of light, c, appears the same to all observers, regardless of their motion, by treating time and space as variables depending on their velocities and relative to the constant c. Measured time varies as the velocity, v, of a coordinate system approaches that of light by a factor of $1/\sqrt{1-v^2/c^2}$. This replaces the classical transformation laws of

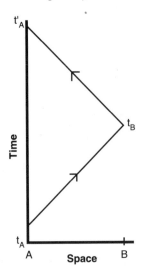

Figure 5 Synchronizing clocks in relative motion

Note: Light travels from A to B and is reflected back.
Clocks at A and B are synchronized if $t_B - t_A = t_A' - t_B$.

vector addition with Lorentz transformations.[44] The time τ for one coordinate system traveling at uniform velocity, v, with respect to another taken to be at rest will be a function of the time t in the second system and vary by a factor of $1/\sqrt{1-v^2/c^2}$, so that $\tau = (t - vx/c^2)/\sqrt{1-v^2/c^2}$, where x is spatial displacement along the x axis. As v approaches c, the factor $1/\sqrt{1-v^2/c^2}$ approaches zero and τ; approaches infinity. Since their measurement is a function of velocity, time and space cannot be completely independent of one another. At the speed of light physical processes would take infinite time. For this reason, and because – due to similar transformations – mass approaches infinity as v approaches c, increasing velocity above c would require infinite energy, nothing can travel at a velocity greater than c. As a result, distant time intervals will be measurable only within limits set by the value of c.

Any event can be located at a space–time point represented by four coordinates (x, y, z, t). From the space–time position of an event E light converges on E and spreads out from E with the geometry of a three-dimensional cone in four-dimensional space–time (Figure 6). Disregarding gravitational effects, every event forms the center in the geometry of an identical light cone. Since c is a constant limiting the velocity of all signals, for every event there will be a light cone marking out a region of those events that can affect E, defining earlier events, and a region of those events that E can affect, defining later events. It is possible for two events to be causally connected if they can be linked by a signal. Given that effects cannot precede their causes, the space–time direction from which light converges on E and travels from E will be the direction from cause to effect.

This clarifies the affinity between physics and B-theory.[45] The theory of relativity defines B-relations, but no A-characteristics. Every event can be represented as the center of a light cone that determines what is earlier and later, but no one event gets privileged as the present moment. The direction of causality, i.e. the direction in which signals travel, determines the direction of time, the B-relations earlier and later.[46] The theory of relativity develops from a critique of Newton's physics, which, ignoring Newton's Neoplatonic leanings,[47] has much in common with B-theory. The principle of relativity preserves Newton's laws but alters the Newtonian time line. Since time is relative to the motion of coordinate systems and the speed of light, relativity transforms the Newtonian geometry of time. For Newton, time is independent of space, and time intervals are the same everywhere. For relativity theory, only integrated space–time intervals remain invariant. The space–time separations between two events are a function of their four-dimensional coordi-

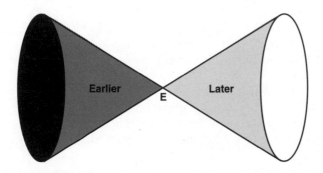

Figure 6: A two-dimensional representation of the light cone for an event E

Note: The boundaries of the cone are formed by signals traveling at c. Event E can be affected by events in the earlier part of the light cone and can affect events in the later part.

nates, such that if c is arbitrarily assigned a value of 1, $s = \sqrt{(x' - x)^2 + (y' - y)^2 + (z' - z)^2 - (t' - t)^2}$. (Note that the difference in sign of the time variable shows that space–time does not treat spatial and temporal dimensions as equivalent.[48]) So, instead of Newton's simple time line, the direction from earlier to later takes the form of a space–time geodesic. This preserves the mathematical conception of time originated by Aristotle.[49] The single B-series comprised of B-relations embodies what is countable in motion with respect to before and after. As for Aristotle, time is relative to motion, now privileging the motion of light and determining the temporal order.

TIME AND EXISTENCE

Time seems closely connected with existence. Events only exist in time, take time, and occur in temporal order. Temporal intervals measure the continuance of things and processes in existence. Things come into existence in temporal succession. Coming to be present and coming into existence often get identified. Evident change raises skeptical questions about the existence of the past and the future.[50] Tense logics treat past, present, and future as modifications of existential operators: to be present, to be past, to be future. Although B-theory rejects pure becoming, it takes earlier and later as real relations between real objects that come into and go out of existence through causal processes.

There is, however, also a reluctance to identify time with existence. Such identification seems to treat time as something substantial, independent of things and events. Instead, time is thought of as a way things and events stand in relations of succession. Accordingly, that present, past, and future are real merely means that something is happening, something has happened, and something will happen.[51] That an event occurs is not the same as it coming into existence. Even some A-theorists insist that becoming present has no existential implications.[52] Moreover, in the four-dimensional geometry of space–time no sense can be attached to enduring or changing other than that certain relations hold between entities.

Some A-theorists argue that not only the four-dimensional geometry of space–time physics, but also tenseless descriptions of past and future events suggest that events stand timelessly or sempiternally in relations of precedence and hence coexist. If events come into and go out of being by becoming present and then past, they cease to exist when they are no longer present. Only events occurring now exist. Consequently, the totality of events cannot coexist.[53] This position takes coexistence to imply the absurdity that the totality of events exist simultaneously. Alternatively, it may be held that coexistence merely indicates that events are related parts belonging to the same whole, not occupying the same moment. B-relations, then, comprise a permanent order of succession, but imply no ontological commitment.[54] The totality of events is neither simultaneous nor timeless, but successive. Moreover, for something to exist must mean more than that it exists now. Otherwise, the claim that only the present exists would be a mere tautology.[55]

Tenseless descriptions of time do not necessitate that temporal realities be eternal, like mathematical facts. Analysts, especially B-theorists, tend to think of the natural world as consisting of particular entities exhibiting various properties and relations. The scope of existential quantifiers formally represents a theoretical commitment to the existence of various entities. Quantification functions to affirm the existence of entities and to count them, but the tenselessness of quantifiers has no implications concerning entities' temporal features. Events, times, or space–time points may all fall within the range of quantification. Even tense may be accommodated: either by interpreting existential quantifiers asserting the disjunction that something exists, has existed, or will exist; or by developing a tense logic treating tenses as sentential operators that may accompany quantifiers and modify their range. This allows existence claims and tenses to vary with respect to one another, ruling out suspect inferences,[56] eliminating, for example, the inference from "X occurred before this utterance" to "X exists," since the former only implies that X has existed.

Since B-series times are countable, B-theory tends to accept quantification over events, times, or space–time points. It amplifies Aristotle's definition of time, transforming what is counted into what falls in the range of quantification, and before and after into the temporal relations earlier and later. It disregards or modifies the Aristotelian association of time with motion in favor of independent or relativistic temporal relations comprising a time line. Moreover, like Locke, B-theory relegates A-characteristics to features of the knowing subject as a matter of perception and pragmatic concerns.[57]

A-theories, on the contrary, do not lend themselves so easily to representing existence claims through quantification. B-relations form a well-ordered series, each member occupying a permanent and unique position excluding all other members. A-theories, however, postulate an indefinite number of A-series, a new A-series for each passing moment. This multitude cannot, itself, form a unified series, since the relations between its members change continuously, and because members fail to occupy a unique position to the exclusion of all other members.[58] This induces McTaggart's argument against the A-series. A-characteristics do not form a series of positions in succession, or segments of a line. Practical discourse and tense logics presuppose some perspective acting as an index that selects a single A-series. But the multiplicity of A-series does not behave well under quantification. If this multiplicity grounds existential claims, as in the thought that only the present exists, then the formal apparatus of quantification cannot adequately represent those claims. McTaggart's argument holds that this vitiates the multiplicity of A-series grounding existential claims.

A-theory may claim either that only the present exists or that the future does not exist. The former relies on the presence of experience.[59] The latter pictures new segments of the universe coming into existence at each moment. Present and past are real but the future is not. This is not merely a version of the moving now. For, rather than some atemporal spotlight scanning along the time line, it proposes that time increases as the universe does and needs no atemporal index. This view adopts Aristotle's notion that claims about the future are neither true nor false.[60] However, if it is now true that the universe will increase its content, and a priori true that the present precedes the future, then some claims about the future are true. Besides, this view has difficulty making sense of causal laws, since they make universal claims about what future events can possibly occur. Causal laws, moreover, ground true predictions about the future given present conditions.

Coming to exist in the present may, alternatively, be identified with the point at which one of various possibilities is realized. This implies no restrictions on the truth values of claims, and is consistent with the causation of events and various entailments constraining the lines of possibility. The past is definitively realized and unalterable, while future possibilities must be grounded in the definite past. The present, then, is a point dividing past, realized possibilities from future, unrealized possibilities. Future things and events will exist and will be specifiable once realized. But now they are merely unrealized possibilities that can only be specified conditionally.[61] That fully realized natural time consists entirely of a

B-series is consistent with this account of the A-series. That the future may follow various possible lines does not nullify the claim that B-relations comprise actual time. Only one of the various possible lines of the future will be realized, and so long as this line is consistent with what is already realized it will form a B-series.[62]

EXPERIENCING TIME

Some asymmetries between and different attitudes toward past and future seem to support temporal passage. That an event will occur or that certain individuals will exist in the future is contingent. The future appears open or unknown to us, but not the past; hence the claim that possibility comprises the future. The future, but not the past, depends on our actions; we may bring about the future but not the past. We feel relief when painful events are past and anxiety regarding those in the future. We await some events, try to prevent others, and dread both preventable and inevitable adversity.[63] B-relations account for temporal direction, but not for legitimate differences in our concern before events as opposed to after them. The fact that a token sentence occurs a particular temporal distance from an event does not explain our concern. A desire to rectify a past error is a wish that it now be earlier than it is, not a desire that our utterance be simultaneous with an earlier date.

These asymmetries show that our ordinary conception of time embraces A-characteristics and temporal passage. However, these asymmetries may only be psychological features that arise out of perception and action. Appeal to such features seems unable to demonstrate whether natural time is transient or the future undetermined. Our ordinary conceptions may well be mistaken, and often conflict with the results of the natural sciences. Drawing ontological conclusions from our ordinary conceptions may express unwarranted anthropocentric biases.[64] Asymmetries in our attitudes toward time do not entail that those asymmetries objectively belong to time. An empirical approach to time, however, cannot stray too far from what appears in experience. Analysts accept the empiricist project of tying all possible reality to experience. Perception constrains any legitimate inferences about features of objective nature. Since time figures in all experience, this approach attempts to understand time from within appearances. Like the Epicureans, this separates perceivable time from conceivable time, and, following Locke, it requires an explanation of the relation between time in nature and our awareness of it.[65]

The most compelling argument for temporal passage affirms the presence of experience. Perception is always of something present; experience itself is always present. It includes an impression of events becoming manifest in the present flowing from the future, and of ourselves flowing into the future. This, of course, also involves experience of events ordered in B-relations. When we observe change we perceive one event preceding another, and we observe the same order regardless of whether the events are past, present, or future. But, while there may be no objective passage among observed events, we clearly have the impression of such passage. Even the staunchest B-theorist acknowledges that the appearance of time in experience exhibits A-characteristics. That one event precedes another must be experienced in the present of some consciousness.[66] Were there no objective time, but only temporal experience, that experience would itself exhibit presence and temporal passage. For A-theory, the presence of experience is evidence for the objective reality of A-characteristics, and the experience of passage for its reality. Since experience is present, and since experiences are events, events must have A-characteristics. It seems absurd to say that in a world without experiences nothing would be past, present, or future. Moreover, even if A-characteristics are merely psychological manifestations or consequences of pragmatic interests, it must be possible to give a coherent account of claims about them and of their production.

The new B-theory distinguishes between the pragmatic meaning of claims about the presence of experience and what makes them true or false. It denies that there are any objective A-characteristics, and without them this explanation for the presence of experience fails. What makes a token of a present-tense sentence true is that it occurs at the same B-series time as what it is about. Experiences must be present, then, since any true token sentence asserting the claim that, or token question asking whether, an experience is present must occur simultaneously with that experience. The combination of the facts that a painful event happened before this utterance, that although we remember the pain our contemporary experience is not painful, that we are relieved and that relief occurs after the pain, and that the relief is an effect of the cessation of pain explains how expressions of relief that a painful event is over can be true without there being any objective A-characteristics. If these facts obtain, then our utterance will be true, whatever additional meaning it may accrue.[67]

Beyond justifying claims about the presence of experience, we must make sense of the experiences themselves. We experience passage and experience itself is present. B-theory explains these experiences by appealing to their causes. What appears in the experience of time need not coincide with what produces these appearances. The experience of presence arises, then, from the causal efficacy of perception. Sense experience results from the interaction of stimuli and our sense organs. Sensory awareness of the world, then, will be simultaneous with the effects of such stimuli. Consequently, the experience of passage originates from the accumulation of traces left in memory by sense experience.[68] Such accounts, however, offer little more than a promise concerning future results in the natural sciences.

CHAPTER XI

Phenomenology of time

> All impression is double: half enveloped in the object, and half produced in ourself.
>
> Marcel Proust

In 1905 Edmund Husserl (1859–1938) gave a series of lectures on the phenomenology of time consciousness. These lectures, modified in ensuing years,[1] contain the nucleus of attempts adequately to describe appearing time as appearing. Husserl thought the problems regarding such descriptions among the most difficult and important in phenomenology.[2] For the consciousness of immanent time embraces all lived experiences, combining them in a unity.[3]

BEYOND NATURALISM

Phenomenology aspires to a more rigorous empiricism than grounds the natural sciences. It eschews inferences purporting to discover objects and laws that transcend experience.[4] It refrains from taking the natural world, objects and events unified by causal laws, as its domain of inquiry. Instead, phenomenology investigates appearances just as they appear, abstaining from assertions concerning existence beyond appearances.[5] Phenomenological investigation, accordingly, sets aside the questions that beset the inheritors of McTaggart's problem. It abjures problems concerning whether time is objectively real, whether time consists of an A-series or a B-series, whether the A-series is subjective or objective, and whether the perception of passage results from external causes acting on our sense organs. Analysts differ regarding these issues, yet agree that time, as experienced, appears as an ordered flow, regardless of how it objectively exists in nature.[6] Phenomenology examines and seeks adequate descriptions of the appearances of time as it is experienced. While descriptions of these appearances must note that some temporal objects appear earlier than others and that some are present and others past or future, these aspects do not suffice to describe time consciousness or appearances of temporal objects.

Phenomenological analysis of time, then, begins by setting aside all claims regarding whatever transcends experience. This excludes assertions about the existence of objective time. A fortiori, it excludes the claims of natural psychology, which explains experience by appealing to causal laws governing psycho-physical organisms. Instead, phenomenological analysis examines time only as it appears immanent in experience. An order of connections inherent in experience appears there that may differ from that of objective nature.

Time, given solely as it appears in experience rather than as an object of perception, is neither objective nor in objective time. Apprehension of objective time presents time as the object of acts of apprehending. These acts fall within the immanent domain of experienced time. Objects of apprehension share a common measure and so belong to an objective order, but it is only through the apprehending of that order that it can be constituted in experience. Phenomenological analysis of time seeks to grasp and describe the variety and connections among temporal apprehensions. Grasp of the contents of temporal apprehensions does not imply a grasp of objective time, since these contents do not compose objective time. Objective time is constituted in experience by the apprehending of these contents as objects of a nexus of apprehendings.[7] Phenomenological analysis examines objective time only as given through apprehensions and only as it is apprehended, intuited, or thought conceptually. Objective time is thus given as the object of various apprehendings.

In addition to seeking to exhibit the apprehendings and acts that constitute time in experience, phenomenology seeks a priori laws that govern this constitution.[8] Again this differs from natural psychology investigating causal laws governing empirical subjects. Phenomenological inquiry can discover a priori laws only as what is invariant in objects of actual or possible intuition. The method always begins with those matters actually given in reflective experience. It tries to discover what may vary in those matters without their changing sense. Phenomenology therefore recognizes that experience is intuitively given as meaningful. It refrains from accepting the theoretical and causal presuppositions of naturalistic empiricisms, which assume that raw experience is meaningless in itself and acquires meaning only through an inferential process of interpretation. Simply to see a chair as a chair requires neither inferences nor additional interpretation. The chair can be given, i.e. intuited; what is given also includes constituents of the chair's meaning or sense. In this way also, phenomenology claims to be a more radically empirical field than that of the natural sciences. For it recognizes that not only particular sensations and objects may be given in intuition, but the non-sensible and ideal as well.[9]

Natural scientific approaches attempt to explain both temporal order and the perception of time and events causally. They explain the direction of time as that from cause to effect, and the perception of time as the result of causal forces acting on an organism's sensory processes. This reliance on causal explanation presupposes realities transcending what is given in intuition. Consequently, phenomenology not only abstains from these sorts of solutions, but also reconceives the fundamental problem of time and time consciousness, focusing on its unity.

Even naturalism acknowledges that we intuit temporal spans. Phenomenology asks, moreover, how it is possible for us to experience a temporal span as unified. We hear a melody not as isolated tones or as chords occurring in a specious present, but as a manifest unity over a temporally differentiated span. It might be thought that the unity of an event is a function of causal relations connecting its parts, and since experiences are events their unity can be explained similarly. But even were this so it would not show how the sense of the events becomes constituted in consciousness – how, for instance, tones in a melody can be apprehended as just past or yet to come. The unity of events over time is not equivalent to the unity of their apprehension. Naturalistic accounts that appeal to the accumulation of effects on the brain omit the experience of a temporal span, since they fail to show how that experience is both unified and includes a perspective from which parts of the span are apprehended as past and future. Apprehending a temporal span is not to have that apprehension caused by some earlier or simultaneous event. These causal relations could hold in sequence without an event ever being apprehended as a unity, since nothing may bind the apprehensions together.[10] Events can be grasped only when their temporal parts are apprehended together as occurring at different times.

Similar considerations apply to naturalistic psychological accounts. Franz Brentano (1838–1917) sought to explain the unity of temporal experience through the association of memorial with perceptual representations.[11] The sequence of tones in a melody causes a sequence of perceptual representations of each tone. Each of these, in turn, includes the content of previous tone representations represented as past: after one tone has ceased a representation of it remains combined with the succeeding tone. Without some trace of the earlier tone remaining we could not experience melodic relations among the successive tones. Yet unless these traces modify the originally presented tones we would perceive these tones as simultaneous. These traces preserve the contents of sensations but modify their temporal determinations. The same difficulties apply to consciousness of the duration and succession of sensations. Some modification is necessary since the endurance of a stimulus merely implies that the sensation endures, not that it is sensed as enduring; the duration of a sensation is not the same as a sensation of duration. It is possible that a sensation could endure without our being conscious of its endurance, either because, once experienced, sensations disappear from consciousness or because they appear simultaneously rather than successively. So Brentano suggested that a series of representations attaches to the original presentation of a tone in sensation, each continuously reproducing the original sensation and modifying it as more and more past. The stimulus produces the representation of a sensation in phantasy, but phantasy itself produces modifications of the representation as past.

Brentano's theory still depends on causal forces operating on a psycho-physical organism to explain the experience of time. For it only needs its appeal to phantasy in order to account for the discrepancy between the causes of sensations and their contents. But by resorting to phantasy Brentano must, like B-theorists, conclude that the apprehension of temporal flux is illusory. Perception only grasps reality in the present moment. Pastness and futurity modify the real, severing their connection with objective reality. As perceptions become characterized as past, they continuously transform the reality of the present into the phantasy of the past. Phantasies of pastness and futurity attach in the present to every passing from and coming into being. Consequently, the reality of times other than the present transcends any possible experience.

Brentano's theory proposes explanatory psychological laws and regards experiences as objects belonging to objective time. Yet, even excluding these naturalistic presuppositions, Brentano's theory fails as a phenomenological analysis, because it fails to account for the fact that duration and succession appear to a unified consciousness.

It fails to show how intuitive presentations can shade into phanta-sized representations. It does not clearly distinguish perceiving a succession from remembering one, consciousness of something temporal but long past from phantasies of something temporal, and representations of past successions combined in current conscious-ness from those that are not. It fails to distinguish between representation, apprehending act, content of apprehension, and apprehended object; consequently, it fails to say to which of these aspects of apprehension temporal moments belong. Primarily, however, it fails because something present cannot become past merely by being associated with a quality of pastness. The quality of pastness would itself have to be present, while what is past tran-scends the present moment. Absurdly, the presence of pastness in the intuition of time makes the past simultaneous with the present. Finally, merely adding a psychic moment cannot transform some-thing real into something no longer real. Therefore the continuous addition of qualities cannot adequately analyze intuitions of dura-tion and succession.[12]

INTENTIONALITY AND TIME CONSCIOUSNESS

By setting aside consideration of anything transcending lived expe-rience, describing only what is immanent to it, phenomenology discovers intentionality, that every consciousness is a consciousness of something: every perceiving, representing, presuming, judging, expecting, hoping, loving, etc. is directed toward something perceived, represented, presumed, etc. Intentionality is the struc-ture, inherent in every lived experience, of directedness toward something. Both psychological processes and naturally existing objects transcend appearances, and hence intentionality is not a relation between such transcendencies. Instead, every intending consists of an articulated but indissoluble whole, of which an apprehending act and an apprehended object comprise abstract parts. Every intending includes some way in which it is directed toward its intended object. Every intended object is given through some mode or modes of apprehension, and its being so given belongs, through the intending, to the intended, regardless of the objective existence of the object. Since it does not transcend lived experience, the intended object need not objectively exist; even a hallucinatory perceiving is intrinsically intentional, directed toward something hallucinated in the perceiving.

160

In a perception of a chair, the chair itself, neither an image nor a representation, is given, but only and precisely as it is perceived, i.e. what is given is the chair insofar as it is perceived. While at any time only one side or surface facing me may be presented to me, I see one and the same chair as I change my orientation to it and am presented with a continuous series of profiles shading off into one another. The content presented to perception in these adumbrations differs, but the perceived remains the same. The chair is given with the sense that it transcends experience. The transcendence abandoned by applying phenomenological method gets recovered in the sense discovered immanent in experience. Perceiving is not simply sensing, but cognizance of structures that can be read off what is given. A simple perceiving of the chair as it is discovered in perception includes perceiving that the chair is there in its entirety (not just the facing surface), is for sitting, was manufactured, etc.

Not every aspect of the chair, however, can be given through every mode of apprehension. The chair is given through a perceiving as the mode of its apprehension, but it may also be given through a picturing, imagining, loving, sitting in, or some other cognitive, affective, or conative mode. In an actual perceiving of a chair, the chair itself is intuited in its bodily presence. A thing perceived may be given in its bodily presence, but something may be given in other modes of apprehension without being bodily present. In perceiving a picture, for instance, the picture is bodily present, but what the picture intends usually is not. Every intending of something can be intuitively fulfilled in a manner appropriate to its mode of apprehension. So, for instance, trying to remember someone's face intends their face, but that memorial intending may be fulfilled by bringing their image before the mind's eye. Similarly, discussing the rules of chess intends the rules, while making moves according to the rules fulfills that intending by behaving in accord with them.[13]

Intentionality is an essential property of all lived experience, so whatever is given in experience must be given through some mode of apprehension and belongs to experience only as so given. Whatever appears must appear as an intending of something, as an abstraction from some intending of something, or as constituted by the synthesis of various intendings or their abstract moments. Brentano's theory of time consciousness failed clearly to distinguish apprehendings from their contents and take sufficient notice of modes of apprehending.[14] An adequate description of time consciousness must show how it is constituted as a unity immanent in experience. Such a description will not restrict itself to temporal contents of experience, but will also examine the temporality of their modes of apprehension.

A temporal object, whether immanent or transcendent, retains its unity while extending over a time interval. The perception of a temporal object also retains its unity while extending over time. The tones in a melody occur successively and each endures; these durations are immanent temporal objects. The intuiting of that succession and the intuiting of each enduring tone also extend over time; neither occurs at a single now. As we have seen, it is unsatisfactory to attribute the apprehension of elapsed tones or phases of a single tone to memory, since that reduces the reality of the perception to a single point joined to memory and expectation. Temporal appearances consist, fundamentally, in the unity of intendings that encompasses temporal objects.[15] When a tone sounds we are conscious of its beginning and intend the tone as present for as long as it continues to sound. We are also conscious of the tone continuing. So long as any temporal phase is present we intend the continuity of its phases as immediately past. We intend each phase, in turn, as more and more past. We also intend the whole duration of the tone as continuously receding into the past. Even when the tone ceases we are still conscious of the tone and its whole duration as having just sounded, but no longer continuing to sound.[16]

This describes the manner in which the enduring tone is given to consciousness rather than the tone itself or its duration. It is one and the same tone and its duration which are given as present, continuing, and past. Focusing only on what is immanent, the tone begins, continues, and ends, and its duration recedes into the past as a unity. The whole duration recedes from the present in which it is originally given; the tone keeps its temporal extent. The enduring tone remains the same, though its manner of appearance differs. Descriptions of immanent objects differ from those of their manner of appearing. We may describe immanent objects which endure and have phases, e.g. the tone's duration which retains its identity as it is initially given and continuously sinks into the past. But this is distinct from describing the manner in which we are conscious of the different phases of the duration of the immanent tone appearing, as perceived by or retained in consciousness. Although the intended object remains the same, the now point is continually new. Acts through which that object is apprehended undergo continual modification.

The tone is generated at the actually present now, and only at that point do we have a primal impression of it. This primal impressional phase changes continuously into something that has been.[17] These elapsed phases, themselves, are retained in consciousness of what has just been. Retentional consciousness is a distinctive mode of intending – a specific manner of apprehension and style of content – that intuits temporal objects as temporal. Retention is a genuine intuiting of the past tone itself, not a construct from sense contents and memorial apprehensions. The retention of a tone is not the perception of an actually present sense content nor the quasi-perceiving of some actually present substitute for a sense content, an image, echo, after-image, etc. One and the same tone is given in retention as was given in perception. But the tone as given in retention differs essentially from that given in perception, since retention is directed toward the tone as past. The perception of an object positions it as now present. However, when the perception ceases, the consciousness of it does not. It is, instead, modified from an impressional to a retentional consciousness. While the object and its duration remain the same, the way they are given varies continuously.

Since immanent temporal objects must be given through retentions and present impressions they exhibit characteristics or "modes of temporal orientation," e.g. being past.[18] The continuity of these orientations, or running-off phenomenon, exists immanently only as an inseparable unity that cannot be broken into segments or phases except as abstractions from that continuity. The form of this continuity does not vary. Running-off characters shade off continuously, analogously to visual profiles shading off into one another when one moves around a spatial object, except that temporal profiles can never be repeated. Once a phase occurs it cannot reoccur. Each beginning phase of an immanent temporal object exhibits the character of being present, each later phase a continuity of pasts. These phases consist of both a continuity of modifications of nows and a continuity of uniform recession of these points into the past (see Figure 7).

Each perception positions the perceived as existing now. When perception ceases it has no now impression, but only retained phases, phases retained in consciousness as just past. Moreover, every retention becomes retentionally modified and retained as more past. Hence, not only does each impressional consciousness pass over into a new retentional consciousness, but these retentions themselves get retentionally modified. A continuity of retentions trail behind each now point, which, as it recedes into the past,

Contemporary traditions

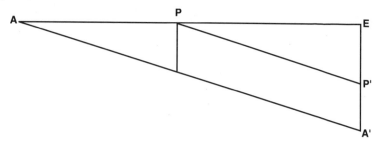

Figure 7: Husserl's diagram of time

Source: Husserl 1991: 29

Note: "AE – The series of now points. AA' – sinking into the past. EA' – Continuum of phases (Now point with horizon of the past)."PP' – sinking into the past.

preserves its sense as having been that now point with those retentions adumbrated behind it. Perception intends the now, but also intends its release as having just been displaced by a new now.

Primal impressions spring into consciousness spontaneously without being produced by it. Consciousness develops these by retaining them in continuously augmented intensities. Since retentions are directed toward something once generated in a primal impression, but no longer existing, it is a priori necessary that a primal impression precede every retention. Individual phases, including the primal impressional phase, can only be given as an abstract limit of a continuum of phases. Moreover, since each phase is only an ideal cut from a continuum, for it to appear in consciousness every present impressional moment must be retentionally modified. The temporal field includes the continuous modification of present impressions and trailing retentions. The whole temporal field is limited and always has the same temporal extent. Temporal objects and their durations recede from the now point in a sort of temporal perspective, where the further they recede toward the temporal horizon, the less distinctly their articulation is given and the more their temporal form becomes obscured. At the horizon, retention of elapsed phases recedes into empty retentional consciousness, pushed back continuously until it finally disappears.[19]

In one sense, to be given in perception consists only of primal impression in contrast with retention. In this sense, perception continuously shades off into non-perception; the now is continuously mediated by what is not now. Such perception, however, is merely an abstract ideal limit toward which the continuum converges. Such abstractions must be founded on the continuum. An adequate perception of a temporal object, then, is a complex act that includes impressional apprehensions of the now, together with continuous retentional apprehensions of the past and their further retentional modifications. In this sense, the object is perceived so long as new impressional phases are given. Retention, in such cases, is perception of the just past given directly to retentional intuition.[20]

RECOLLECTION AND TIME
CONSCIOUSNESS

While the just past is given to retention in intuition, it is re-presented in recollection. Simple recollection may occur vaguely in a flash, but recollective acts may also be performed clearly in reproduction. Every originally presenting consciousness can, in principle, be re-presented precisely. Recollective acts are themselves apprehendings given in impressional consciousness as re-presenting something, occupying its own temporal position, and exhibiting running-off phenomena. Yet the objects reproductively given in recollections do not occupy the same temporal positions as the acts that apprehend them. What is reproduced in recollection shares a temporal structure analogous to that of perception. When the hearing of a tone is recollected, the recollected now corresponds to that of the perception. A temporal fringe of recollected retentions trails behind this point, in which earlier phases are re-presented. Recollection, then, differs from perception. For something now ongoing is, itself, presented to perception, while recollection re-presents it. The recollected now is not itself presented, but only reproduced as having been originally presented.

Since each now is replaced by a new now that sinks into the past, what was given in perception determines intendings aimed at the past. These intendings can be fulfilled by recollective leaps into the past, intuitively re-presenting the past up to now. Something still living in retention, e.g. an object's identity, may likewise be re-presented. Recollections position their objects as having been perceived, but are fulfilled by immanent reproductive intuitions of those objects. Time consciousness can become evident only when reproductive and retentional flows coincide. Recollection can be fulfilled in reproductive re-presentation of a succession given in a still active retention. What gets represented intuitively in recollection is the objects themselves given as having been present. Recollections need not represent the earlier perceiving of those objects, although reflective recollection of such perceivings is always possible as well.

Nor do recollective representations imply the construction of an image or other internal object that merely resembles or corresponds to the original intentional object. Re-presentation intends the original itself as having been present and as now past. As re-presented, the original is positioned with respect to the present as absent, but as having been present at an earlier now. Reproductive positioning inserts the reproduced in immanent time as reproduced. Even if subsequent experience contradicts what is thus objectively positioned, the recollection is retained as it occurred.[21]

Thus recollection constitutes an enduring object. The continuum of retentions that attaches to the recollected now exhibits the character of being past, but need not be related to the actually present now through a continuous series of appearances. Recollection re-presents the continuum of phases in a reproductively modified quasi-perceiving, in which the recalled object is given as if it were present, although no such object is actually perceived.[22]

In re-presenting enduring objects, recollection intends a unified continuum that includes not only a now and its retained phases, but also intendings aimed at the future relative to that now. These intendings, or protentions, are aware of what actually followed that now, but only insofar as what followed had already been anticipated, and anticipated as to some extent problematic. The distinction between what has been, as given in retention, and what has not yet been, as given in protentions, is no abstraction. Whether anything at all will be given is purely contingent, and belongs to the possible, not to what has actually occurred.[23]

Every perception belongs to a nexus of more or less distinct intentions, inseparable from a protentional as well as a retentional horizon. Recollection re-presents this entire temporal context. Intendings directed toward the temporal context, which are not the same as the complex of intentions aimed at the enduring object, position a duration in that context. Every original constitution of an enduring object protends its future as an indeterminate possibility that may arise. Recollections directed toward the future of the re-presented now, however, can be completely determined by what had already been perceived following that now. An indeterminate empty intention aimed at the possible objective series of events in time, then, always belongs to the reproduced past. This temporal horizon surrounds actually recollected durations with intentions that insert them into time.[24] This temporal horizon accounts for consciousness of an A-series; recollective apprehending provides the dimension of re-presentation necessary to avoid the paradox. The insertion of recollections into time generates neither a vicious circle, since the inserting apprehendings are distinct re-presentations, nor an infinite regress, since no recollective apprehending can be repeated, although its content may be.[25]

The succession of one content after another is itself given as an additional element besides the succeeding contents. One content emerges, is retained, then another. We are conscious of each content, the interval between them, and that the second succeeded the first. Consciousness of successive contents intuits the succession itself.[26] A recollection of succession is not a succession of re-presentations. We can freely repeat the succession itself and any consciousness of it. Likewise, consciousness of duration can only be recalled more or less adequately, not itself given again.

Recollection contrasts with retention. Unlike recollection, retention is neither reproductive nor representational. Retentions point back to original presentations in primal impressions, and we can simply attend to retentional phases without producing them anew. Retention holds in consciousness, as just past, what has been constituted in impressional consciousness. Recollection, in contrast, reproduces intentional objects that endure through a continuum of retentions. Such reproduction does not explicate something persisting in consciousness, but something already temporally modified.[27] Besides, acts of recollection themselves occur over temporal spans, while retention can occur only in continuously new phases. Reproductions can also be re-presented freely; they can be brought before consciousness more quickly or slowly, distinctly or confusedly, as a whole or in parts. Retentions are never optional but can occur only passively. They can never be repeated or varied. Retention never re-presents its objects; it always must have its intended object before it. Furthermore, retention and recollection have their own modes of evidence and obscurity. Retention gives its object with absolute certainty. Since the succession of the temporal flow is given reproductively in recollection, it is possible that the order of the reproduced flow differs from the actual order of succession. As already noted, recollection can be contradicted by further experience. So reproduction may add its own varieties of obscurity to whatever errors and obscurities may occur in original apprehensions.[28]

Phantasy is also a mode of re-presentation. Something can be re-presented as now existing but neither present nor remembered, so long as there is some possible route by which an impression of the object can be obtained. Since in phantasy the intended object is not itself given, it must always be of the same kind as something already given. Yet not every re-presentation is phantasy. Recollection, in particular, is not. For in phantasy a now has no determinate position in the temporal field, while in recollection what is reproduced is always given as actually occupying a position with respect to the present, as coinciding with a past time. Regardless of its intuitive content, phantasy apprehends re-presentational contents modified to exclude position; these contents stand in no determinate temporal relation to the contents of perception. There is never a continuous transition from perception into phantasy because no primal impression can arise without a continuum of retentional modifications, the whole of which is actually positioned. This is why Brentano's theory failed.

CONSTITUTING IMMANENT CONTENT

Various modes of time consciousness belong to a unified temporal field that gradually comes to givenness. Temporal unities are originally constituted in the continuum of protentions, impressions, and retentions. In addition, time consciousness re-presents these unities in recollection, expectation, and phantasy. Temporal unities arise in original impressions, re-presentational modifications in reproductive consciousness. Re-presentation is always derivative. The content of primal impressions are first generated at their now points, while re-presented contents are given as rising to the surface once again.[29] The temporal field is not a chain of associations. Instead, new actualities arise, already protending what is to come while fulfilling previous protentions. Everything new produces a retroactive effect that necessarily extends back over the whole temporal field, continuously modifying past appearances.

For any unity to appear to consciousness it must be constituted as a unity, i.e. it must be given in a series of actual and possible intendings through which we become aware of it and its unity. All appearances, transcendent or immanent, must be constituted in the temporal field. Everything transcendent must be given through something immanent, and immanent objects must be constituted in the flow of temporal phases. Immanent contents are adequately given, while transcendencies are not, since they can never be given completely, but only through adumbrations of a series of profiles.[30] Recollections do not originally constitute immanent contents, but only reproduce them as already constituted in the flow of temporal phases.

Immanent contents include apprehendings of transcendent and immanent objects, cognitive, affective, and conative intentional acts, sensations, and appearances, heeded or not. Intending of something perceived can be directed either toward the content as it arises and runs off, immanent in the perception, or toward what becomes constituted as a unity that remains identical in that flow, immanent, but not part of the perception or the flow. Immanent in perception, what is perceived continually runs off regardless of differences in content. So the flow of immanent time itself can be distinguished from the flow of its contents. What maintains its identity in the flow manifests a distinction between something identical in the flow and its changing phenomenological properties. Both these identities and the identities of their abstract moments must be constituted as such in the flow. This constitution resembles that of transcendent objects, in that it unifies various profiles. However, these identities are completely and adequately given through their temporal profiles.[31]

In either case, adequately given immanent contents must have temporal extent. Immanent contents preserve their identities only insofar as, in their actual duration, they are retained in prior phases and protended in coming phases. The duration of every immanent content occupies a specific temporal position. In recollection, each duration must be positioned in time by intentions aimed at its temporal context. Determinate intendings flow away into indeterminate intendings delimiting a horizon. As reproduced, the duration must be explicated with the unity of the time stream through these indeterminate retentions and protentions. The complex of intendings directed toward what is reproduced can be fulfilled by the system of appearances that belongs to that object. But those intendings aimed at the temporal context can be fulfilled only by a continuity of connections tying the object to the actual present. Every re-presentation is itself an intending originally presented through some immanent content. Recollective intendings therefore arise and are presented as a series of intendings with their own horizon that cannot be isolated from the horizon of their recollected contents.[32]

Since the order of immanent connections may differ from the order of connections in objective nature, the duration of perception need not be equivalent to the duration of perceived natural objects.

In nature, perception and perceived, therefore, need not be simultaneous. Phenomenologically, transcendent objects may exist before and after they are perceived. Immanently, it may seem as if impressional consciousness must be formed before it can be given to consciousness. As originally constituted, however, retention and consciousness of what is immanent are perception, inextricably united with primal impression. So described, immanent perception and perceived must be simultaneous.[33]

Simultaneity is constituted when primal impressions arise at a single now point. Continuous immanent production of primal moments always forms a multiplicity which undergoes temporal modification as it runs off. Although each primal multiplicity runs off as a simultaneous unity, all belong to a single flow, always sharing the form of transience from not-yet to no-longer now. Primal impressions share the same running-off modes, even though they may vary in duration and in other intentional modifications included in their content. The being-together-now of this complex does not vary, but gets retentionally modified as a whole in a continuum of phases. This retentional being-together unites with the primal impression complex, constituting an enduring nexus of intentionalities, positioning each point in a single flow. This intentional nexus unites the multiplicity of impressional consciousness and the complex of its running-off modes; hence the constitution of simultaneity cannot be extricated from that of temporal succession.[34] Immanent time is constituted as a unity, a being-together of immanent objects and contents, necessarily and continually elapsing as ever new primal unities arise. The constituting of this unity of simultaneity and succession can no longer be appropriately described with temporal language.[35]

Since we are conscious of the flow of immanent time, it too must be given in some mode of intending: the unity of the flow of time consciousness, the unity of simultaneity and succession, must itself be constituted in the flow. Consequently, the flow of consciousness must constitute its own unity. This is possible through the double intentionality of retention.

Retention holds on to both variable immanent contents and variable modes of their apprehension, thereby constituting the unity of immanent objects. New phases, with their contents, displace previous ones, constituting interrelated unities appearing as phases of a flow. Through these unities, retention also helps constitute objective unities immanently given, including the unity of objective time.[36]

However, rather than aiming through the flow toward what is intended in it, reflection can also aim at the flow itself. Through retention, reflection may intend immanent objects, since apprehendings get retained as well as apprehended objects. Since retention is itself a mode of apprehension, each phase of the flow grasps the continuously preceding phases, including their retentions. Each retention of a retention also retains what is retained in the preceding retention, ultimately continuing back to the retention of a primal impression. Each phase of the flow grasps the being-together of a continuity of retentions of preceding phases, which is itself continuously modified; each new continuity is retention with respect to the total continuity of the preceding phase. This continuity of retentions constitutes the unity of the flow as immanent time. It constitutes a unity of elapsed retentions of the flow as itself an immanent object adumbrated in the flow.[37]

Hence the constitution of immanent time requires a retentional grasp of the being-together of elapsed phases and its limit at the point of primal impression. As new phases arise and push back retained phases, this being-together continuously becomes retentionally modified. As new moments arise, retentional being-together continuously coincides with itself in a non-voluntary synthesis, identifying already retained being-together as a part of a newly retained being-together, thus constituting a unified flow. Since this unified flow includes continuously arising now phases and protentions of not yet actual phases, it comprises a partially actual order, unified in a quasi-temporal arrangement of phases by the pre-phenomenal, pre-immanent synthesis of time constituting consciousness itself.[38]

The phases of the flow in which its own phases become constituted cannot be identical with those constituted phases, which are necessarily already retained past phases. Although they differ, constituting and constituted flows coincide. Constituted phases continuously become re-identified in their being-together with newly arising phases. In this way, not merely a sequence of static durations, but a unified continuous flow appears. Constitutional reflection retains a being-together, which, unified with the continuous addition of something new, sinks into the past. Since this flow constitutes itself as a phenomenon in itself, no additional flow is necessary; the flow becomes apprehended in its own flowing.

EXCURSUS: CONTINUITY AND PRESENCE

Recent arguments indicate the possibility of a break in temporal continuity between primal impression and retention. They object that the evidential value of perception is at odds with the continuous constitution of time consciousness in retention. These criticisms claim that phenomenology presupposes that presence possesses a privileged evidential value, and that, to the extent that perception blends with non-perception, its evidential value decreases. The criticism argues that it is only possible to compare perception and reproduction, original presentation and re-presentation, retention and recollection because the perceived present compounds presence and non-presence, perception and non-perception. Retention, on this view, is a form of non-perception, of re-presentation, inextricably united with punctually present perception. Since no perception is possible without retention, all perception becomes re-presentation, posing a threat to phenomenology's rigorous empiricism.[39] To maintain the evidential value of presence would require a break between the present instant and its retention. For, were retention, as non-perception, to share evidential status with primal impression, perception would reduce to re-presentation.

Husserl distinguishes perception as apprehension given now in primal impression from perception as the act originally constituting temporal objects through the continuous unity of temporal apprehensions.[40] It is in contrast to perception as primal impression that the criticism considers retention to be a form of non-perception and so of re-presentation. This, however, confuses temporal presence with givenness. Perception contrasts with retention only in the limited sense that retention and protention are different modes of givenness from being given now. Something can be given to consciousness as fulfilling an intending without it being temporally present. Elapsed phases of a temporal object are still given in a modified way – as just past. Retention is itself a mode of intentionality, with fulfilling intuitions peculiar to it. Nothing could possibly be retained without first arising in the now phase. But retention does not make that elapsed phase into an object which it looks back on; it retains what was given in the impressional phase, making it possible to look back on the object through a reflective glance.[41] The now phase can only itself become an object of retention in reflection. Yet retention is not necessarily reflective consciousness; nor is it a repetition or duplication of what is given in primal impression: it retains or holds on to exactly what arises in primal impression, positioning it as being past. Retention transcends its actual present phase of consciousness to grasp something actually past, without transforming what is past into something present.[42]

Unified experiences of temporal spans are possible, then, because givenness extends beyond the present instant. Perception, as an act of original constitution, must present a unified temporal span. Any re-presentation of something perceived intends what was perceived as having been presented in that span. Recollections compare with retentions since both intend something as past, but they do not position their objects the same way. Recollections position the whole temporal span as having been perceived at some earlier time continuous, but not necessarily immediately contiguous, with the position of the recollective act. Retentional, ongoing, and protentional phases posit their objects as immediately contiguous, belonging to a continuum of phases. These phases are modes of appearance distinct from what appears through them. When something new arises, it comes before consciousness through the now phase. The now phase orients temporal experience and, unlike what appears through it, does not recede into the past. Instead, it links just past and immediately to come phases.

Aristotle's conviction that time cannot be composed of nows also holds true for time consciousness.[43] The now point of time consciousness can be approached only as a limit. Although actual continuation of time consciousness is contingent, discontinuity is not possible at every point; it presupposes continuity. Time consciousness parallels Aristotle's contention that the now both joins time intervals and divides them. However, what is joined is not alike, in that the actual is joined with the potential, retention of the actual as past with protention of the possible as future; the now connects retentional with protentional consciousness in a continuum. Unlike in Aristotle's view, the now does not limit change in something else, but limits the phases of actual consciousness as they open onto future possibilities, and limits protentional phases as they are actualized.

CHAPTER XII

Transcendence and existence

He walks on and gradually unrolls the uninterrupted ribbon of his own passage, not a series of irrational unrelated images, but a smooth band where each element immediately takes its place in the web, even the most fortuitous, even those that might at first seem absurd or threatening or anachronistic or deceptive; they fall into place in good order, one beside the other, and the ribbon extends without flaw or excess, in time with the regular speed of his footsteps.

Alain Robbe-Grillet

Time consciousness unifies a continuously arising flow of intentionalities to be together. The continuum of temporal positions flows homogeneously from the generative now, each position continuously modified to become more and more past. Temporal positions constituted in the flow adhere to an essentially transitive, asymmetric, and irreflexive order. Each position may be filled with a variety of intentionalities, in which complex impressions may occupy the same position while varying in content. The possibility that positions, themselves, become objectivated is founded on temporal modifications of these complex impressions.[1]

The temporal flow functions as a transcendental condition for anything at all appearing in experience, including the temporal flow itself. As in Kant's transcendental deduction, time functions to unify experience and its objects. But, rather than by representing the past and future through transcendental imagining,[2] the temporal flow unifies through the being-together of intuitive temporal intendings. The flowing being-together of this nexus of intentionalities acts as a necessary condition for the constitution of any possible experience of unities, whether transcendent, purely immanent, or the unity of time consciousness itself. This transcendental function, already implicit in the analysis of time consciousness, discloses questions there about the ways these unities exist.

CONSTITUTING TRANSCENDENT TEMPORAL OBJECTS

For transcendencies to become constituted in the temporal flow, they must simultaneously be given through immanent contents. Multiple temporal adumbrations, united through the synthesis of temporal apprehension, constitute immanent unities.[3] The same impressional consciousness involved in constituting immanent unities also functions in the constitution of transcendent unities, but not through the same intendings. By intending immanent unities as appearances of something transcendent, immanent unities can also constitute transcendencies. Accordingly, the constitution of transcendent unities is always distinct from that of the immanent unities in which their appearances are constituted.

For a temporal object to appear in the immanent flow as one and the same, it must be possible for that identical object to be freely repeated in recollection. Acts of recollection re-present the identical object as had already been intended. That indefinitely many such acts be able to reproduce and re-identify the same object is necessary to constitute that object in time.[4] Each such act coincides with the others and the original perception in identifying the same object. When two recollections intend the same full intentional context, including the same temporal horizon, they recall the same temporal object. Because recollections also repeat an object's temporal horizon, successive recollections position it continually further in the past.

The natural world can only appear on the condition that physical things are constituted in and through the same impressional consciousness in which their perception becomes constituted. So, unified immanent apprehensions and their contents becoming constituted in the original flow of experience comprise necessary conditions of any consciousness of physical things. Physical things can appear through immanent apprehensions and their contents by way of higher-level intendings of those apprehensions with their contents. For, continuous adumbrations and multiple presentations dispersed in the flow of experience can present one and the same object. Immanent contents are intended as appearances of physical things when they are intended as appearances of parts of unities that can be perceived through different types of perceptual apprehendings: those that endure through changing orientations and adumbrated spatial profiles; those that always have aspects not now appearing yet which can in principle be repeatedly perceived; those that appear now only in some of their aspects but cannot appear at once as a whole; and those that can possibly appear to others.[5]

Physical things not only become constituted with their durations, but also retain their durations and positions in a single objective temporal extension. Many physical things can occupy identical points in this extension. Hence, it must be possible for objective time to be constituted as a unity over and above the temporal objects that occupy it. Objective time is intended as something that can be presented only in part but never as a whole. Like physical things, it appears through immanent unities that posit it as a unity appearing in changing orientations. However, unlike particular temporal objects, objective time has no limits, neither beginning nor end-point. Objective time, then, appears as the temporal form of appearing objects because appearances of objective time are themselves presented immanently with their own duration and position.

The constitution of objective time functions as a transcendental condition for consciousness of any transcendent temporal object or event. Immanent contents, in turn, provide transcendental conditions for objective time to appear. That objective time becomes constituted in consciousness requires that durations and positions of objects can be fixed, identified, and reproduced in different intendings. Yet the now phases are not preserved, but run off into the past. It must be possible, then, to contrast fixed temporal positions with the flow of time consciousness. Objects seem both to change their place in time and to preserve it. Temporal objects and positions do not change their place in time, only their distance from the actual present, which is a continuously new point in objective time.[6] This problem pertains to both transcendent and immanent temporal objects.

Temporal objects become constituted as unities because the same object can appear through many immanent contents. Because the immanent contents through which the same temporal object appears present it with the same sense, a variety of apprehendings can identify it, even though it lasts for some duration, which may have elapsed, and even though its apprehension may occur discontinuously. In each phase of the apprehension of a temporal object the identical sense must coincide partially with the sense of the object given in its other phases. The sense of physical things, for instance, presents them as unified but indefinitely complex centers of causality.[7] To appear as having their typical forms of behavior, physical things must preserve their durations and temporal positions.

Objective time becomes constituted because the modifications that comprise the flow do not belong to the sense of things, immanent or transcendent, that appear in the flow. Even when the appearances of a temporal object remain identical over a span of time, their modes of givenness undergo continuous modification. Appearances, given as now, becoming retained, and sinking into the past, are posited as appearances of the same thing. Even perfectly similar immanent contents belong to different actually present now points.[8] Each phase of a duration includes a consciousness of itself as subject to temporal modifications, and so as belonging to a unity of duration. This apperception preserves the position and duration of temporal objects by positing them as identical in the flow of temporal modifications. Temporal modifications require neither new objects nor new time points; nor do they alter their intended objects. They modify identical objects together with their durations.

Individual time points, their temporal positions and durations, become constituted with their unity and identity because an identical now point with its content becomes positioned in time as it sinks into the past. The same temporal individual maintains its position and duration; only its mode of givenness changes from now to continually more past. Temporal positions and their contents continuously recede into the past because new primal being arises in each new now. Retention modifies each new content together with its temporal position to receive the characteristic of being past. This, in turn, continually becomes modified, increasing the temporal distance between the retained position and each new now. Every apprehended actual temporal position must be apprehended as bounded by an ever new now. It is necessary, then, that primal being and modified being have different temporal positions. Two appearing phases have the same position only if they spring from the same now point, and necessarily different positions if they do not.[9]

Each now phase differs individually from every other now phase, even if their impressional content is perfectly alike. Each new now point individuates the content that arises at that point. So its arising at a new temporal position distinguishes one primal impression from another regardless of their content: primal impressions arise, presenting ever new time points. Since these impressions arise continuously, temporal positions run off continuously. These positions differ from primal impressions because they are continuously modified. The continuity of temporal positions arises because temporal positions originate in a continuous moment of individuation that constitutes the identity retained as they become past. Continuous individuation arises because primal impressions continually differ from their modified contents. This individuation adds nothing to what is individuated. That is why temporal positions can be modified together with their contents. That a primal impression arises as an individual identity is sufficient to guarantee the identity of the temporal position that runs off. In this sense, individuation is the form of time in which actual contents arise at the now point.

The flow of consciousness is ineluctably temporal, but not everything that becomes constituted in consciousness is temporal. Neither mathematical matters, nor propositions, nor values exhibit temporal properties. Particular states of affairs are not strictly temporal, although they are founded on perceptual presentations of something temporal. However, we must distinguish these cases from the acts of judgment through which they are given. Acts of judgment become constituted temporally, while the propositions that become constituted as the objects of such acts have no temporal features. Judging unifies a process of positing that ends in a belief regarding the proposition that has come to consciousness in it. Judging not only holds onto in retention what arises in primal impression, but also constitutes a syntactical complexity posited in what arises. The intending of this syntactical complexity begins, continues, and ends with the judging, but the syntactic structures exhibit no temporal features.[10]

Physical things become constituted in experience through the flowing off of their appearances, along with and through immanent unities in the flow of impressions. Appearances of transcendent objects become constituted because immanent apprehensions and their contents become constituted. We have already seen how immanent apprehensions and their contents become constituted as unities through the unified flow of immanent time.[11] A physical thing becomes constituted as a temporal unity because the continuity of apprehensions unifies both the thing's changing and its dispersed appearances as belonging to something enduring. Hence, physical things can only be given in experience through the temporal flow; their constitution presupposes the constitution of time. Immanent temporal modifications constitute objective time. Ultimately, the unified flow of immanent time will be self-constituting.

CONSTITUTING THE ABSOLUTE
TEMPORAL FLOW

The contrast between flowing time and fixed temporal positions was McTaggart's concern.[12] As McTaggart acknowledged, appearances change, and the paradox arises, regardless of whether time is real.[13] We experience both B-relations and A-characteristics. Phenomenology shows how both can appear in experience. Fixed temporal positions of intentional objects bear B-relations, while flowing modes of givenness exhibit A-characteristics. Because B-relations and A-characteristics do not apply to the same appearances, McTaggart's problem does not arise for phenomenology. The separation of intentional objects from their modes of givenness avoids a vicious circle, but may still seem to generate an infinite regress. Whether temporal modes of givenness (protentions, primal impressions, and retentions) generate a vicious infinite regress turns on how they become constituted as unities in consciousness.

Transcendencies experienced in nature become constituted as unities through immanent contents and in multiple appearances. The time of physical things corresponds point for point to the immanent time of their appearances.[14] This immanent time includes all intentionalities, embracing their apprehensions and contents. Intentionalities themselves become constituted in the flow as enduring unities standing in B-relations. Immanent objects are enduring unities in immanent time, constituted by passively occurring synthetic functions, continuous identifying and distinguishing syntheses, which comprise the flow rather than any unities in it. Both transcendent and immanent unities are distinct from the synthetic functions through which they are constituted.[15] Enduring individual unities may change, and change at varying rates with respect to the same duration.

What appears at a single time point, however, must be a phase of an enduring or changing process.[16] Temporal modes of givenness make the constitution of immanent time embodying B-relations possible. Such modes of givenness can only belong to the phases of immanent time as part of the meanings intended in the continuum of retentional consciousness. The flow is a unity of phases in which enduring unities in time become constituted. This constituting flow is distinct from the phenomena it constitutes. Phases are not processes; nor do they persist or maintain their identity over time.[17] Each phase exemplifies its own mode of givenness, including its many ways of being given in manifold retentions and protentions. Although each phase preserves previous phases, those phases are only given through modifications as increasingly past. Since a single phase has no temporal extension, it does not endure, change, or vary in rate.

Yet everything that appears must appear through experiences occurring in the flow of immanent time. This must include the temporal modes of givenness that constitute immanent time itself. The duration of a temporal object, its unity, is originally constituted by its being given in a continuity of primal impression and retentions. These modes of givenness are also given as a unity. Since each apprehension is simultaneous with what it apprehends, the unity of apprehensions through which a temporal object is given occurs in exactly the same duration as its object. This unity becomes objectivated[18] with its position through passive continuous identifying and distinguishing syntheses, and like any object can be re-presented in recollection.

Insofar as phenomenology makes sense of A-characteristics by means of temporal modes of givenness, it may seem to generate an infinite regress, producing a variation of McTaggart's problem. If temporal modes of givenness must themselves be positioned in immanent time, and this must be constituted in a unity of temporal apprehendings, and apprehensions themselves must be positioned in immanent time, then a vicious infinite regress similar to McTaggart's seems to follow. It seems impossible for any of these apprehensions to be given. Since each new intending must be part of the flow, the flow as a whole seems unable to become the object of an intending.

Husserl avoids this supposed regress through describing an absolute time constituting consciousness.[19] It is absolute in the sense that it includes everything immanent in experience, all intentionalities, apprehendings, and their objects as given through them. This consciousness constitutes immanent time, but can do so only insofar as what is impressionally given can be integrated into a flow exhibiting the invariable form of an impressional now phase shading into a continuum of retentions, on one side, and a horizon of protentions, on the other. Absolute time constituting consciousness flows, in that temporal modes of givenness necessarily change continuously without a permanent content or substantial unity.

Absolute time constituting consciousness provides a necessary condition for all constitution. But it is not, itself, a constituted unity. While it is possible, in reflection, to take unitary parts of the flow as objects, these are not unities of something that persists. Objects must endure and something must endure through a process, but, since phases of the flow cannot persist, the flow itself can be neither object nor process.[20] Instead, the absolute flow is the individuating form of any enduring or changing being. Therefore it exhibits no temporal features. It can have neither duration nor temporal position, so it cannot exist in time, or in a temporal relation to anything that does exist in time. The absolute flow has no temporal properties and cannot be the subject of temporal predicates; even calling it a flow is metaphorical. We can only describe it as arising at the primal source point, and in terms of what actually becomes constituted.[21]

Although the flow of time consciousness makes sense of A-characterstics, it does not form a series. Members of a series must each occupy one position in the series to the exclusion of all others, and they must maintain constant reciprocal relations.[22] In contrast, as temporal apprehensions become positioned in the flow their intended now points successively occupy the actually present now point. Their intended now points continually change their positions relative to the actual now point as they recede into the past. Yet, through these changing relations, immanent time becomes constituted as a single continuous temporal field. This field unites all phases of the flow in a continuous synthesis that identifies the time of each apprehension as it runs off with a position in the same duration as all other apprehensions.

When objectivated and interpreted, the absolute temporal flow is necessarily identified with the time of apprehensions and their contents.[23] For the absolute flow to escape the regress, it must be able to perform continuous identifying and distinguishing syntheses without falling into paradox. A consciousness of the absolute flow must be possible without the flow itself simultaneously being objectivated. Reflection on constituting temporal phases is possible because these phases are already given in primal impression. The present impressional moment is a limit, but its actuality can only be grasped in reflection on the retention of its content, including its temporal position. Retention is possible only for something already given in primal being. Consequently, primal impression must include the whole continuity of immanent time as being-together all at once. The unity of duration must arise at the now point without being constituted by various intendings. There must be an awareness, simultaneous with the actually present now point, of the whole of time consciousness as a temporally extended continuity of phases.

Awareness of apprehensions does not necessarily turn toward them, objectivate them, or position them in time. Before turning toward them, there already are experiences which appear as phenomena when turned toward, differentiated, and individuated as the object of some intending. The turning toward differs from its object, even though that object was already there to become the object of regard. A further reflection can distinguish this difference by noticing that the act of turning toward the object supervenes on something previously unregarded.[24] After the object is turned toward in reflection it becomes differentiated by its being grasped through the reflective glance. Everything intended appears against a background of what is already there to be grasped. But it must always be possible, in principle, for this background to be objectivated in some actual intending. In this way, the absolute flow is self-constituting; it needs no additional flow but becomes apprehended in its own flowing.[25]

Primal consciousness is not an apprehending act.[26] It involves an awareness of the intentionalities that arise in it that does not attend to itself. Every primal being is conscious in itself, without being objectivated. Within primal being there already lives an awareness of protentions and retentions that does not yet take them as objects of reflection. This awareness lives in primal being similarly to the way that while a new phase emerges retention retains the preceding phase but does not attend to it as its object. Just as there is an apperception of the identity of an object within an intending of it, there is an apperception of the unity of absolute time constituting consciousness already in the actually present now phase.

EXISTENCE AND TEMPORALITY

The disclosure of this apperception within primal being points back to the continuous identifying syntheses and presages the transformation from a phenomenological psychology into a transcendental philosophy. The conditions that must hold for intentionality to be possible cannot, themselves, originally be objectivated. The temporality already manifest in primal being provides the condition necessary for the unity of time and of appearances in time to be possible. By unifying intentionalities, temporality provides the transcendental condition for all appearances. Absolute time constituting consciousness provides a unifying perspective directed toward a temporal span. This perspective, as already contained in the arrangement of phases in primal passive synthesis, does not itself exhibit temporal properties. A continuous identifying synthesis cannot be positioned in time. It has no predecessors, successors, or tense. It is not an event in the succession of events. What has arisen in it passes, but it cannot pass. It is a differentiation and ordering of moments, not a changing series. It is the original source from which all that appears arises. It exists as constituting the flow of appearances, and as constituting its own appearance in that flow.

This peculiar way of existing raises the ontological question of what it means to be. This question does not ask which things exist, but asks for a conceptual understanding of existence as such. Martin Heidegger (1889–1976) explores this question by offering an existential interpretation of phenomenology; an interpretation already implicit in Husserl's work.[27] Heidegger explicates it by interpreting appearances as beings, and investigating the ontological difference between being as such and particular beings, distinguishing different ways of being. Heidegger argues that different aspects of time ground this ontological difference.[28] In particular, an existential interpretation of the distinction between absolute time constituting consciousness and the entities it constitutes grounds the difference.

Since, fundamentally, phenomenology analyzes intentionality, Heidegger focuses on the analysis of that entity whose existence is characterized by intentionality, by its comportment towards beings – the *Dasein*. "*Dasein*" is one of the German words for existence; it means "'to be here." Heidegger uses the term to avoid any suggestion that intentionality is a feature of the traditional ego or subject. Intentionality is not immanent in subjectivity, but presupposed to it. Since intentionality is neither subjective nor objective, nor a relation between pre-existent subjects and objects, *Dasein* is neither a mental substance, nor isolable from the world. In its comportment towards objects, *Dasein* exists in the world, the meaningful world as it is given in experience. *Dasein*'s comportments are always ways of being in the world. Hence, Heidegger interprets the existential ground of intentionality, *Dasein*'s way of being, as being-in-the-world.

Being is equivocal; different sorts of beings exist in different ways. Intentional objects are given primarily in the context of their use. Beings encountered in this context exist as functional entities whose sense is determined by their functional role therein. Their mode of being, defined by their functional sense, is to be handy.[29] (A hammer is handy insofar as it is used for hammering, and it is at-hand insofar as properties such as its shape and composition are observed.) The mode of being of intentional objects, given as what *Dasein* merely finds before it, abstracted from but available to perform functional roles, is to be at-hand.[30] These modes of being are ways entities that do not themselves possess intentionality can be said to be within the world. They exhibit within the world the sense imparted from their functional roles for handy things, or from what properties they exhibit for entities at-hand. In contrast, *Dasein*, as such, exists neither as handy nor at-hand. *Dasein*'s way of being is to be intentional, to be directed toward, i.e. engaged with and concerned about, the world.[31] *Dasein* ordinarily concerns itself with handy or at-hand entities, but it can also concern itself with other *Dasein*s, with itself, or with the modes of being exemplified by any of these entities. This ability to be concerned with beings and their modes of being identifies *Dasein*'s being-in-the-world.

Dasein's mode of being differs radically from that of intraworldly entities. However, this ontological difference and the meaning of being-in-the-world become clear only with the exposition of their relation to time and temporality. Time conditions what it means for handy and at-hand entities to be. *Dasein*'s temporality makes possible its concern about such entities and their modes of being. More importantly, *Dasein*'s temporality makes possible its concern about its own being and mode of being. This enables *Dasein* to ask after the meaning of its own being. The ability to ask this question distinguishes *Dasein* from other kinds of beings. Analysis of *Dasein*'s mode of being by way of temporality attempts to clarify the ontological difference and account for its existential unity.

Heidegger recapitulates Husserl's analysis of time consciousness in different terms, emphasizing its existential and pragmatic aspects. The transcendental attitude refrains from taking appearances as part of a world with aspects that cannot appear. Heidegger's hermeneutic method interprets appearances as beings showing themselves, and consciousness as the being-in-the-world of *Dasein*. He begins by making the everyday concept of time problematic, in order to show that it presupposes temporality as a condition of time's appearance and unity. He reads Aristotle as having articulated the everyday concept of time, then argues that Aristotle's analysis requires temporality as its condition of possibility.

Heidegger considers the now to be crucial to Aristotle's analysis of time as what is counted in change with respect to before and after.[32] When following the transition in change, we are able to keep track of how many nows have passed by repeatedly saying "now" to mark the transition. The now indexes the before and after in counting change. It is itself in continuous transition, and so is able to mark the transition in change. Transition has the formal structure of a continuous unified stretch from something toward something. In experiencing change we retain the former and expect the latter.[33] Time counts the before and after of any possible change, fixing its limits and so measuring its duration. This formal continuous stretch is a necessary condition of any possible phenomenon of change.[34] The now binds the before and after together, holding time together as a continuity in transition. The now unifies that from which the change occurs with that toward which it changes. Since the now looks forward and backward it cannot be correlated with a point, but rather contains time as a stretch within the now. Transition itself occurs within the now.

Time, then, cannot be a sequence of nows, since in each now time stretches into non-being on both sides. Time, then, cannot merely be an abstract succession of immobile parts.[35] Since each point is itself a transition in the continuum of the flux of time, the image of a line misrepresents time. Hence it is misleading to see the now merely as a limit. Because it is essentially in transition, it is open on to earlier and later nows; the now can mark a limit, but it is not itself a limit.[36] In marking a limit the now is counted, but the now is not a limit insofar as it undergoes transition. The now in transition between earlier and later comprises a horizon in which it is possible for changes to show themselves. This horizon embraces all possible changes, and all beings.

Aristotle claimed that things are in time the way what is counted is in number.[37] Time is not dependent on the intrinsic content or the mode of being of change. For something to be in time means its transitional character can be embraced by time. The horizon of the now in transition between earlier and later embraces all phenomena, all beings that show themselves. To be in time, then, means to be embraced by this horizon. Consequently, time is not itself one of the beings in time. Time embraces all change and so must always already be given for any change to show itself. In this sense, time is an a priori horizon, holding around things, in which they can be given in successive order.[38]

For time and change to show themselves nows must be countable. Yet nows can be counted only when followed through change. To follow and count nows requires that they exist within time's a priori horizon, which embraces beings and holds them together in the sequence of nows from earlier to later. This adapts Aristotle's claim that nothing can be counted except by the soul.[39] But, rather than entrench the subject–object dichotomy, Heidegger examines the problems of what time is and how it exists by examining how *Dasein* exists. For *Dasein* encounters and comports itself toward beings. It encounters beings only if they show themselves within a meaningful world, and such a world must be embraced by time.

Aristotle conceptualized the common understanding of time as a sequence of nows. When counted, the sequence of nows is unlimited, unidirectional, and irreversible. This can become accessible because the common understanding of time is somehow already directed toward time; this common understanding points to temporality, the condition of its accessibility. We comport ourselves toward time in practical orientations.[40] We concern ourselves, for example, with how much time a task takes or how much time remains to do something. Although we do not focus our concern on time, we must reckon with time in order to accomplish tasks. *Dasein*'s primary comportment toward time, then, is to use time in guiding its actions.

Time is not something at-hand that *Dasein* directs itself toward; rather, time shows itself insofar as *Dasein* guides its activities in accord with the time they take. In taking time, *Dasein* already has time without reifying it. The common understanding of time is explicit in measuring practices, but it is also expressed implicitly in each of *Dasein*'s comportments. *Dasein* exists as taking time and expresses its temporality in every one of its comportments. *Dasein* does not comport itself toward the now as an object. Rather, by implicitly saying "now," *Dasein* expresses its comportment toward something present. Comporting toward something in *Dasein*'s present enprésents it: *Dasein* orients itself toward that thing as present. Likewise, by implicitly saying "not-yet-now" and "no-longer-now," *Dasein* expresses its comportments toward something future or past, respectively, expecting or retaining it. In its practical engagements *Dasein* reckons with the unity of past, present, and future horizons.

In (implicitly) saying "now," "not-yet-now," and "no-longer-now," *Dasein* expresses the temporality of its comportments. Enprésenting, expecting, and retaining interpret the sequence of nows as a complex structure of the world-time belonging to *Dasein*. *Dasein* engages with handy things with respect to their functional significance. The totality of such relations of practical engagement affords *Dasein* a significant world, a world that exists for the sake of various ends. Within this world, different times will bear different significance depending on how those times matter. Times may be appropriate or not with respect to fulfilling various functions. Consequently, different times will neither be abstract nor float free from what occurs at those times. Each time will be datable as the time at which some specific event is occurring. Moreover, time's stretch between before and after varies in the significance of its durations. The tasks and events that date a time, as when some event is occurring, also structure the span through which the now endures. As expressed, the now is not a point, but a significant span sharing the duration of some occurrence. Finally, expressed time is intrinsically public. The same dated duration of an occurrence may be expressed by any *Dasein*.[41]

Expressing its temporal comportments, *Dasein* is itself always in transition. It is already oriented in transition towards what is not-yet-now, and from having had in its present what is no-longer-now. In the former case *Dasein* exists as expecting what is later than now, in the latter case as retaining what is earlier than now. Enprésenting, expecting, and retaining are essential moments of the now as transitional, as embracing beings, and making it possible for nows to be counted, and so for anything to be in time. Enprésenting is implicit in every expecting and retaining. Every expecting of something expects it from a present, and every retaining of something retains it for a present.[42] By being-in-the-world *Dasein* exists temporally as the unity of expecting, retaining, and enprésenting.

ECSTASIS AND TEMPORAL UNITY

Dasein derives the time with which it reckons as well as the sequence of nows from its own temporality. When *Dasein* expresses its comportment toward a being or occurrence, it temporalizes it as not-yet-now, no-longer-now, or now. *Dasein* exists as comporting itself towards these temporalized objects of its concern. What occupies *Dasein* is presented to it but need not be temporally present. Only what *Dasein* enprésents is temporally present. Things expected are future, things retained past.

Whatever is expected, retained, or enprésented, *Dasein* is expected, retained, or enprésented along with it. *Dasein*'s comportments, hence *Dasein*'s existence, are possible because *Dasein* exists outside itself, carried away from itself toward what occupies it. *Dasein* understands itself by expecting its own possibilities and retaining what it has already been, along with the way it now finds itself. Hence, *Dasein* exists outside its own present, carried away to its own future and past. It exists temporally as the unity of its coming toward itself from its own possibilities, going back to its own past, which constrains its possibilities, and dwelling with its own enprésenting, in which its possibilities are realized.

Heidegger calls these modes of temporality "ecstatic" because they stand outside themselves.[43] Expecting, retaining, and enprésenting are ecstases, in that each carries itself away toward something. The unity of these ecstases, temporality,[44] carries itself away to other beings, and temporalizes them to be at different times, not merely in the present. Temporal ecstases are intrinsically open onto a horizon in which beings and occurrences manifest themselves. Each ecstasis opens onto a horizon that determines a changing boundary on the manifestation of beings. Ecstatic-horizonal temporality, the unity of temporal ecstases, then, bounds the appearing of beings in the horizons of future, past, and present.

Ecstatic-horizonal temporality, then, makes *Dasein*'s being-in-the-world possible. For the ecstases go beyond themselves by themselves. *Dasein*'s being-in-the-world is primarily characterized by its concern for the matters which occupy it. *Dasein*'s practical concerns deal with matters future and past. In these concerns, *Dasein* exists by dwelling amid what occupies it. Hence, ecstatic temporality makes intentionality possible, makes it possible for *Dasein* to dwell among what occupies it. This differentiates *Dasein*'s mode of being from that of things handy and at-hand. Such things exist in time, but they are not themselves ecstatic. Although *Dasein* may be taken to be something handy or at-hand, this still presupposes that it is so taken by something with an ecstatic mode of being, a *Dasein* in its primary mode of existence. *Dasein* exists, fundamentally, as a temporal entity, i.e. as ontologically constituted by ecstatic-horizonal temporality.[45]

Ecstatic-horizonal temporality also makes possible the structural constituents of expressed time, from which the sequence of nows is abstracted. Every expecting, retaining, and enprésenting is related ecstatically to some being or occurrence, making it possible for that being or occurrence to be encountered. Each enprésenting expresses itself by saying "now," and so points to some being encountered at that time, which can index and so date that time. The unity of ecstatic-horizonal temporality is already stretched over time, because temporality is intrinsically outside itself in the horizons of future and past. Because the now expresses the unity of enprésenting, expecting, and retaining, each now expresses a temporal span, concomitant with a datable event. Each now already spans a stretch of time. Since the horizon of ecstatic temporality bounds anything temporal that *Dasein* can encounter, everything manifest in time must be spanned by that horizon. Each now manifests a transition because it expresses the temporally differentiated span of ecstatic unity. Furthermore, since ecstatic temporality is intrinsically outside itself, its unity is already open on to the world accessible to any *Dasein*. Hence each expressed now is publicly accessible.

Since ecstatic temporality makes *Dasein*'s being-in-the-world possible and the world *Dasein* occupies itself with is always significant, expressed time must already have significance.[46] All comportments toward beings – cognitive, conative, and affective – involve an understanding of what it means for them to be. This sort of understanding makes it possible for beings to be encountered in comporting toward them. It need not be conceptual. Since *Dasein* comports itself toward beings, understanding is a mode of *Dasein*'s existence.[47] In every comportment *Dasein* finds itself situated where something already matters. Such encounters are possible only in unity with *Dasein*'s understanding, because understanding grasps possibilities inherent in what is actually encountered.

Understanding is a projection of something on to its possibilities, without which the understood could have no significance for *Dasein*. A being can be understood only provided it is projected on its own possibilities of being. These possibilities are neither handy nor at-hand, but are possible ways for some *Dasein* to dwell with that particular being. A piece of equipment can be understood only through the possibilities for its functioning. So, for instance, a hammer makes sense as used for hammering in order to construct something. The being of such equipment, as handy, is thus constituted by its individuation in its functional context, which comprises the possibilities necessary for the equipment to be what it is and function as it does. *Dasein* understands the being of equipment in using it, and thus has already projected it on to its functionality, its specific way of functioning.

Entities that are handy have significance through their functionality, which is structured by "in-order-to," and similar functional relations.[48] These relations make sense of equipment and constitute its functionality. *Dasein* understands equipment because it already understands functionality as world, as the horizon in which it can encounter intraworldly beings. As a horizon, world is no more something functional than *Dasein* is one more thing in the world. World is the horizon in which that which is handy and at-hand appear. World is neither something handy nor something at-hand, but belongs to *Dasein*'s mode of being. *Dasein* exists out beyond itself, understanding itself from its world.

Dasein expects that for-which something functions, retains that with-which it functions, and enprésents it as a specific piece of equipment exemplifying functional relations. Functional contexts, then, are structured by temporal relations, and understanding what it means for something to be handy presupposes ecstatic temporality. *Dasein* is able to understand the world of functional significance, because ecstatic temporality opens onto that world which is part of *Dasein*'s being-in-the-world and so shares *Dasein*'s mode of being. Ecstatic temporality, then, makes *Dasein*'s understanding of its being possible.

In its purposeful engagement with practical matters it is essential to *Dasein*'s existence also to be occupied with its own ability to be in the world. *Dasein*'s understanding consists in its projection of itself on its own possibilities. Since all comportments involve projection, projection is not mere contemplation of an empty possibility, but *Dasein*'s engagement with its own possibilities. Projection projects the particular being, *Dasein*, on its own ability-to-be and, with this, the being of other entities it encounters. Since understanding is projection onto a possibility, *Dasein* understands its own existence in its projection on its ability-to-be. Heidegger's most telling characterization of *Dasein* is as that being for whom its own being is an issue. *Dasein* exists as a nexus of intentionalities which is able to understand its own possibilities of existence.[49]

Regardless of whether *Dasein* understands itself from its concern with things or its concern with its own ability to be, *Dasein* exists as the unification of its possibilities with its actuality. *Dasein* exists by projecting itself on its own possibilities, understanding itself in understanding its possibilities by way of its actuality. *Dasein* exists as coming toward itself from its own possibilities, i.e. possibilities of how it can be, given what it has been, and how it at present finds itself. Since *Dasein*'s possibilities are part of its being, *Dasein* is never wholly present; its being-in-the-world presupposes its temporal openness. *Dasein*'s being, then, could be completely actualized only were it no longer to have any possibilities – no future. Of course, for *Dasein* this point can only draw nearer asymptotically.

Dasein's understanding is the way it lives in its comportments without necessarily being comported toward them. *Dasein*'s unity, therefore, is given through its possibilities; i.e. so long as it exists, what *Dasein* has been and is now belongs to what it will have been and so will belong to any possible world it will be able to exist in.

Since the ecstatic-horizonal unity of temporality makes *Dasein*'s being-in-the-world possible and understanding is a mode of *Dasein*'s existence, temporality makes *Dasein*'s understanding of its own being possible, and thereby its peculiar unity.

Dasein's self-understanding is usually neither explicit nor thematic in its concern. *Dasein* understands itself through its engagement with things and, through them, with other *Dasein*s.[50] *Dasein* implicitly understands its world in its significance. Its ways of comporting itself are ways of explicating its understanding of the world's significance. In understanding its world *Dasein* understands itself as dwelling among things in the world. *Dasein* projects its own ability to be on the feasibility, urgency, expediency, etc. of those things in the world that occupy it. This projection keeps *Dasein*'s interests turned primarily toward those intraworldly things in which it is already interested. This conceals *Dasein*'s mode of being and its temporality from it. Consequently, *Dasein* interprets its temporality as something at-hand – the sequence of nows. This interpretation isolates the sequence, abstracting it from its expressive structure. It treats time as one more thing within the world. This interpretation generates both Aristotle's argument for the unreality of time and McTaggart's problem.[51] So, for example, despite the finitude of *Dasein*'s temporality, *Dasein* deduces the infinity of the sequence of nows. For it interprets the sequence as something at-hand where each now is always preceded and succeeded by another now isolated from the conditions of its manifestation.

This interpretation is an unavoidable consequence of *Dasein*'s ecstatic mode of being. But it only partly conceals time's expressive structures, which show themselves as transition, embracing change, and that time is counted. These handy aspects of time point back to the moments of temporality but cannot make them explicit. Through its ecstatically temporal way of being *Dasein* necessarily has an understanding to which being is given and, despite its usual concern with beings, is able to explicate the meaning of being. Temporality makes possible *Dasein*'s understanding of what it means to be, transgressing its usual concern with beings, because temporality consists in the projection of possibilities as such. *Dasein* can make temporality an explicit object of its concern, because it can project its own being onto time in comporting itself toward its own possibilities, understanding its own temporality without comporting itself toward temporality. So the ontological difference between being and beings is latent in *Dasein*'s existence as comporting itself toward beings, and can become explicitly and conceptually grasped.[52] To make temporality conceptually explicit is to step back and interpret its moments according to how they can be understood, i.e. according to the horizon in which they appear and the projection of temporality on that horizon.

The functional context of something handy gives its possibility. Even something missing or broken is possibly handy, even if it happens to be unavailable for use. To miss something is also to be comported toward it, to expect it as needed for use. The temporal horizon, retentive-expectant-enprésenting, which makes sense of the temporality of being handy, whether available or not, Heidegger calls praesens. Both the presence and absence of something in a functional context are modifications of praesens. Something handy can only be encountered in an enprésenting if its handiness is already intelligible. Praesens is the condition of the possibility of understanding handiness as such. Understanding projects beings on praesens as the horizon in which something handy can become manifest. Praesens' structure of presence and absence makes understanding of everything encountered in an enprésenting possible.

Praesens is not the same as the now, for the now is in time. The now presupposes praesens as time's horizon. Nor is praesens the same as the enprésenting of something in time, since praesens is not itself ecstatic. Nor is praesens subjectivity, since it can be understood by *Dasein* as that horizon it is carried away toward. Praesens is the horizonal schema of the ecstasis of enprésenting onto which comportments are projected. The ecstases, however, make the projection of comportments possible. Each ecstasis projects itself on a horizon. Enprésenting ecstatically projects itself onto praesens as its schematic horizon out beyond itself upon which it projects itself. Praesens is the horizonal structure of the present which completes the temporal structure of enprésenting: enprésenting projects itself on praesens. The other ecstases also have their respective horizons.[53]

These horizonal schemata comprise the outermost periphery within which beings can be present or absent, that on which the possibility of any being, its being, is projected. Heidegger interprets ever wider horizons of being, terminating in the temporal horizons upon which being is projected – horizons not belonging to being. Being, then, can be understood from these horizonal schemata upon which temporal ecstases project themselves. Heidegger interprets progressively prior projections terminating in the temporal ecstases. Since the ecstases consist in self-projection as such, being can be projected upon the horizonal schemata. Since it projects itself, temporality makes understanding of being and comportment toward beings possible. Temporality projects possibilities onto the temporal horizon, being onto time. It is the source of possibilities as such, the a priori.[54] So ecstatic temporality is a condition of the possibility of all projection, and so of the horizon of anything actual or possible on which and toward which projection takes place.

Enprésenting removes everything encountered, carrying it over and projecting it onto praesens, on which it opens. Each ecstatic carrying-away-toward intrinsically opens onto its horizon and keeps that horizon open. Projection, as such, is possible because the ecstases form a unity, each modifying the others.[55] Heidegger sees this unity as primordial temporality, the event of givenness as such, out of which everything emerges. The event or givenness as such[56] includes all projection, its horizon, and all that can appear within its horizon. It is not a being, neither subject nor object, although it makes the unity of beings, including *Dasein*, possible. Something particular always arises out of it, but this arising is always dynamic. For Heidegger, the event is both necessary and mysterious. While the event secures the unity of *Dasein* and of intentionalities, it pushes the need for unity back to temporality itself. The event is the necessary unity presupposed to all being. It opens up a space in which beings and their being can become manifest. Yet it is an arising, not a being, a dynamic emergence that cannot be brought under any higher ordering principle.[57]

CHAPTER XIII

Multiplicity and virtuality

Chaos . . . primary disorder, in the ineffable contradictions of which space, time, light, possibilities, virtualities were still in latency.

Paul Valéry

The tension Heidegger discovers between the dynamic arising of being and the necessity for the unity of the event would have been no surprise to Henri Bergson (1859–1941). Although it is plausible to see Bergson as a forerunner of phenomenology, there are significant differences. Where Heidegger pursues an interpretive unity, Bergson finds pure multiplicity. Bergson claims that time, what he calls *durée*, cannot be captured by the traditional concepts of unity and multiplicity. For Bergson, *durée* is the condition of anything actual and thereby also of possibilities as such. Moreover, where Heidegger finds the unity of primordial temporality in the event of givenness, Bergson pursues the production of that event in the actualization of the virtual. Such actualization is not a higher ordering principle of unification; rather, it is the creation of organization.

PURE MULTIPLICITIES

Bergson distinguishes two types of multiplicity: one quantitative, the other qualitative. Quantitative multiplicities are numerical multiplicities, qualitative multiplicities non-numerical. He defines numbers as syntheses of one and many, i.e. numbers consist of collections of many identical units. For a number to be a multiple of some units those units must be identical to one another. For many identical units to form a collection they also must be somehow distinct from one another. To be distinct from one another while

194

remaining identical they must be both grouped together and separated in an abstract space.[1] Without such an abstract space, units could not be distinguished unless they were qualitatively differentiated, hence abandoning their identity. Moreover, what serves as a unit may only do so provisionally. If a unit can also be conceived as forming a collection of identical units juxtaposed in an abstract space, it will itself be divisible, and such divisibility can proceed indefinitely. Number, Bergson concludes, must be conceived in terms of juxtaposition in an extended, if abstract, space.[2]

A single variable defines this abstract space in which numbers unify manifolds. This variable symbolizes a homogeneous class of objects that differ quantitatively, but not qualitatively. In such a space, what is divided and the divisions thereof are perfectly equivalent in kind. Only differences of degree can occur in such a space. Numerical differences, therefore, can be divided without changing in kind. Additional independent variables may add further dimensions to such a space, but do not alter its homogeneity.[3] Every point must be identifiable by an assignment of values to each independent variable. Extended to any number of variables, such a space can be modeled mathematically by Riemannian geometry. Multiplicities that can occur in Riemannian space are homogeneous, quantitative, and contain only differences of degree.[4]

Bergson extends the notion of multiplicity to qualitative differences. In contrast to quantitative multiplicities exhibiting only homogeneous differences in degree, qualitative multiplicities exhibit heterogeneous differences in kind. Qualitative multiplicities are more specific and concrete than can be given by any abstract general account of the one and the many. While such abstractions are useful, their very abstractness and generality makes them unfit to explain which unity and what multiplicity are involved in concrete differences.[5] Formal precision and analysis seek generality independently of subject matter. Bergson, however, advocates a concrete precision that requires, in addition, adequacy to the qualitative content of the subject matter. Hence Bergson claims that qualitative multiplicities contain even greater precision than quantitative multiplicities.[6] Consequently, qualitative multiplicities are neither manifolds unified in an abstract space nor syntheses of abstract units in collections.

Instead, Bergson conceives of qualitative multiplicities as continuous, containing differences of kind in concrete *durée*. In Riemannian spaces the number of elements contained in discrete multiplicities serves as their common measure. For continuous multiplicities in Riemannian spaces, that which they are used to model provides the value of their metric.[7] Since they exclude qualitative differences in kind, Bergson denies that such spatial models are sufficient for an adequate conception of continuous multiplicities. Instead, continuous multiplicities are qualitative and belong essentially to *durée*. *Durée* divides continuously in successive wholes, instead of juxtaposed parts; it cannot divide without changing in kind. It may be measured only by varying its metrical principle with each division.[8]

Durée forms a whole that cannot change without altering qualitatively; in succession, it contains differences of kind that permeate one another, fused together in a whole rather than juxtaposed in an abstract space. Since *durée* is successive, it divides continuously, changing as a whole, where quantitative multiplicities do not change. Their divisions are fixed, as expressed by independent variables. *Durée* divides continuously, and so is multiple without being associated with number.[9] Since *durée* also continuously changes in kind, it is not a quantitative multiplicity.[10] While *durée* divides continuously, at each successive stage it includes a totality of differences in kind, permeating one another in a fusion. Its flux is not without structure, but its structure is wholly different at each successive stage of division. At any given stage of division, *durée* forms an indivisible whole. But since it never ceases dividing it forms a succession in which the whole at each stage differs from its predecessor.

Continuous quantitative multiplicities differ in degree regarding quantities of the same kind. Since their metric comes from outside, their unities presuppose some qualitative differences. However, while quantitative multiplicities are actual, qualitative multiplicities are virtual. Whatever is actual can be divided by differences of degree without changing in kind. Its differences, whether real or merely possible, are always actually in the object. Everything quantitative, all matter and space, is completely actual; it can be divided indefinitely without a change in the totality of actual or possible qualities perceivable in the object.[11] In contrast, only pure tendencies, not existing actually but solely virtually, in principle, differ in kind. What differs in kind differs only potentially, as a tendency, i.e. virtually. Although the qualitative multiplicity of *durée* is as real as any actual quantitative multiplicity, its reality is purely virtual.

The distinction between the virtual and the actual must be differentiated from that between the possible and the real. Bergson argues that the universe is truly creative; it issues in unforeseeable novelties, which are entirely new, even as possibilities. Possibilities do not pre-exist their realization, but come into being along with the created things that instantiate those possibilities.[12] Reality is not merely a possible combination of stable pre-existent elements. Were that the case, realization of a possibility would necessitate limiting the possibilities to one, plus the addition of reality to that possibility. Everything real would resemble the possible insofar as realization of a possibility would exemplify the concept or image of exactly that possibility and exclude all others. The possible would include more than the real, since not every possibility is realized, but would be less than the real, since no possibility can, qua possibility, be real.[13]

Bergson rejects this picture, first, because he repudiates the contrast between being and nothingness as part of a false problem. The notion of nothingness, he contends, is a misapprehension of the use of the word "nothing" to focus our interest on one thing while suppressing another for which it substitutes. Hence, possibilities cannot be nothings which subsequently become realized, and the possible cannot include less than the real. While there are real differences in kind, these differences neither have the form of alternative possibilities defined, negatively, by their opposition, nor come into being by limiting pre-existent possibilities until one is selected to be realized.

Second, Bergson offers an alternative conception of possibility as the concept or image of a thing cast into the past. Only when something new is created can it have always been possible. So Bergson claims, for instance, that the next work of art is not yet possible; it will have been possible only when completed, and likewise for creations of nature. As an explanation, the realization of the possible must be retrospective. For only in retrospect could there be alternative possibilities such that some other possibility could have occurred. Confusion arises because, while novelties are indeed possible beforehand in the sense that there was in nature no insurmountable barrier to their coming into being, it does not follow from this that novelties were foreseeable or that their concepts precede their creation. The retrospective nature of the possible does not entail that realities can be put into the past or work backwards in time, only that possibility does not precede reality, any more than a mirror image can precede that of which it is an image.[14] While it is true of closed material systems that later states are calculable from earlier states, such systems are abstractions from creative nature as a whole.

In contrast to the relation between the possible and the real, the virtual produces the actual. The virtual is real and efficacious without being actual.[15] The actualization of the virtual comprises the production of everything actual, including both actual realities and possibilities; if something is a possibility, it is an actual possibility. Since possibilities are abstract images of the real projected into the past, there can be no pre-existent possibilities that could then be realized by acquiring existence. For possibilities to be projected into the past, the past must be something real and concrete. Nothing virtual is ever actual, any more than any possibility, qua possibility, is real. The past, then, is real, although entirely virtual. Actual differences do not resemble the virtualities they actualize, the way the real resembles the possible. Instead, in its actualization the virtual continuously creates differences of kind; actualizing itself is this movement of differentiation. *Durée* is the unfolding of novelty, the virtual actualizing itself. Consequently, the virtual is inherently in flux. It is qualitative multiplicity continuously differentiating and creating its own actualization in differences of kind.[16]

DURÉE AND SPATIALIZATION

Bergson distinguishes between the creative flux of *durée* and the analyses of motion found in traditional metaphysics and modern science. Both assume a stable reality beneath time and change. Platonic metaphysics, for example, supposes that what is immutable passes into motion through a decrease in reality. In modern physics, motion possesses its own reality but is governed by immutable mathematical laws. In either case Bergson objects that analysis gives a distorted image of motion by inverting the ontological relation between stability and flux.[17] Such analyses describe motion using concepts signifying already actualized possibilities. Such analyses inevitably fail to grasp the actualization through which novelty unfolds. They treat time either as a series of states spread out in actual juxtaposition or as a unified progression. Any attempt to express both the unity of movement and the numerous juxtaposed states as belonging to the same time results in sacrificing one, or in confusion.[18] These alternatives are McTaggart's A- and B-series, and give rise to his paradox. According to Bergson, both are artifacts of analysis. So Bergson should not be called an A-theorist – *durée* is not an A-series.[19]

Physicists do, of course, measure motion, in which enduring physical objects alter their positions with respect to an abstract time. This time is homogeneous and measurable only by comparing when motion begins and ends to some standard movement, i.e. by comparing simultaneities. This standard defines the equality of two durations, not *durée*. So measurement deals with time and motion by only ignoring *durée*, and treating it in terms of equations for something already completed and, in effect, immutable. Since *durée* is pure multiplicity, only the actual immobile track abstracted from the virtual flux of *durée* can be divided into homogeneous and thereby countable parts.[20] This is the position of B-theory, where quantification over space–time points interprets Aristotle's notion of the countable, and temporal passage is relegated to the subject or to pragmatic concerns.[21] The B-series, then, is actual but abstract. Analysis pulverizes time into an unlimited number of sections, converging, at their limit, on instantaneous mathematical points, each without duration. Moreover, the unity that binds these sections together must be immutable since analysis can only conceive change in terms of intervals between one point and another. Measurable time, then, reduces time to intervals and simultaneities – abstract time unifying immobile sections.[22]

Concrete motion is composed of an actual homogeneous interval covered and a pure transition or virtual heterogeneous act of covering. The successive positions covered are distinct from the pure passage between them. We misconstrue the transit from one point to another, by seeking fixed points where transition occurs, reducing transition to infinitely small intervals without ever capturing passage itself. Without passage, no interval could ever be traversed. Reduction of time to intervals and simultaneities confuses virtual heterogeneous qualitative differences with actual homogeneous intervals.[23] Instead, pure *durée* is heterogeneous and continuous. It contains neither intervals nor juxtaposed equivalencies. When it divides into successive parts it changes qualitatively. Measurement requires repeatable qualitatively invariant intervals to serve as a standard. Since *durée* differs continuously, no such equivalencies endure. Pure *durée* cannot be measured, then, because no identity continues over successive differences. It is in this sense that Bergson says that because durations cannot be superposed on one another *durée* continuously eludes measurement.[24]

Such analyses try to synthesize *durée* from infinitely repeated juxtapositions. Yet each point only marks a fixed place abstracted from genuine flux. Analysis counts abstract halts in the concrete flux of *durée*, isolated invariant snapshots abstracted from concrete variation. Analysis thereby confuses combination with succession.

It misses real succession, finding only juxtaposed instants and their abstract unity. Real motion must occur in *durée*; it cannot be recomposed from these analytical constituents. The moving body only passes through the juxtaposed points; it would be contradictory for motion and immobility to coincide. Our analyses project these points as lying beneath the moving thing, as locations where the thing would have been had it stopped. Hence, when we intuit transition we misconstrue it by assuming that something immobile must underlie flux.[25]

Bergson often repudiates the analysis of *durée* into abstract time and immobile sections, saying it spatializes time. In doing so, he does not utterly reject the representation of time by spatial models. Such analyses result from the fact that we are active creatures whose perceptions select only those features of what is perceived that are important to our needs and projects. Perception filters out all properties of things that are not useful for our virtual actions, i.e. our tendencies to act one way rather than another.[26] Moreover, as social creatures we adapt to social life through the use of symbols and language. Both perception and language, whether ordinary or scientific, abstract from concrete flux, consolidating continually changing objects into objects that are invariable and familiar. This consolidation promotes social life, but it also generates philosophical confusions, mistaking the most useful notion for the clearest.[27]

Language consolidates things under fixed concepts, by selecting features in common with other things. In this way, concepts abstract and generalize, indefinitely overstepping the concrete differences in their objects. This looseness of fit between concepts and objects both allows for alternative conceptions, contingent on which features get highlighted, and forestalls precise representation of *durée*. Common sense and natural science conceive a world convenient for action, thereby misapprehending *durée*. Science represents time by an abstract line in order to facilitate prediction and control. Concrete things never repeat or continue without alteration, while symbols remain useful only by glossing over differences. Symbolism in science and traditional metaphysics immobilizes this alteration, thereby abstracting even from its perception. It represents time negatively, as degenerated eternity or failed coexistence of what are really successive points, and misses the positive features of *durée*.[28]

To discover real *durée*, then, we should not be satisfied with utilitarian concepts or the symbols that express them. Such concepts are imprecise, for they are too general and could apply to a world other than ours.[29] Instead, Bergson urges us to reverse this habit of thought and intuitively follow the fluctuations of *durée* without presuming its unity. His approach does not abandon concepts, but requires that, first, we make the effort intuitively to put ourselves into *durée*, and only then invent concepts to express precisely each problem that arises. Intuition grasps concrete *durée* in its disparate qualitative multiplicity, but in doing so continually upsets concepts, forcing them to fluctuate. The continuity of *durée* implies both multiple differences in kind and their mutual interpenetration. Because of their dependence on our pragmatic needs, neither concepts nor sense perception can reconcile both these differences and their interpenetration.[30]

In spite of this difficulty, Bergson still adheres to an empiricism, for only in experience can we be given real existence. Intuition gives *durée* in its interpenetrating fusion, which is divisible only into qualitative tendencies. Intuition, by its nature, focuses on and repossesses *durée*, which contains all qualitative differences in their pure multiplicity and flux. The totality of scientific observations can enumerate differences of degree, but not exhibit their fusion. In contrast, intuition both is experience and leads beyond experience to its conditions. These conditions, however, are conditions for actual experiences. They are neither general nor abstract conditions for all possible experience (except insofar as the actual is a condition of the possible), as sought by transcendental philosophies. Instead, they are virtual conditions for actual experiences, and cannot apply to anything other than what they condition.[31] The conditioned as experienced in intuition finds its conditions only as tendencies differing in kind and occurring above what Bergson calls "the turn of experience," the point where experience becomes biased by pragmatic considerations.[32] When purified of their pragmatic distortions, such tendencies can be recomposed, beyond the turn, as virtual conditions of actual experiences.

VIRTUAL COEXISTENCE

Bergson first developed his conception of *durée* in terms of consciousness. In his early work he even identified *durée* with consciousness, in which qualitative states are felt. We are conscious of qualitative differences, for example, in the total effect of melodies or strokes of a clock, and in dreams.[33] Conscious states exemplify a perpetual becoming, and the continuous prolongation of the past into the present. While consciousness is not identical with *durée*, it is more than an example. For we intuit *durée* directly in the flux of consciousness. Consciousness can be seized intuitively only as *durée*, in its pure qualitative multiplicity. The notion of a unified ego is only a convenient way of evoking this intuition.[34] Although he never abandoned the notion that consciousness has its *durée*, Bergson increasingly explored *durée* beyond consciousness in unconscious memory, the existence of the past, the *élan vital* of creative evolution, and the universe as a whole.[35] In each case he insisted that all creation and novelty must arise out of *durée*. But even in these later explorations he appealed to the intuitive evidence of consciousness.

Conscious experience contains composites of clearly specifiable, relatively stable aspects regarded as mutually external, together with continuously changing inexpressible aspects organizing themselves in an interpenetrating whole. Every conscious state can appear in either of these aspects. The former predominate when symbolic representations of succession include conscious states as homogeneous elements arranged to extend one another in an aggregate. This aggregate is ordered according to the congruence of conscious states, and their objective causes, with practical needs. Such elements, occupying abstract time, are fragmentary schematic expositions of the whole of consciousness. While consciousness retains some of the mutual externality of these elements, its changing aspects predominate when intuition reflects on consciousness as an interpenetrating qualitatively differentiated whole.

Intuition is neither instinct nor feeling, but a reflective grasp of this interpenetrating qualitatively differentiated consciousness as the whole of *durée* in its passing.[36] Through their habits, needs, actions, and techniques, people fabricate the stabilities necessary for practical life. Because of these practical needs, consciousness artificially breaks up the whole of *durée* into mutually external things. But pragmatic necessities also give rise to factitious problems when they mask and distort intuition, in which consciousness attends to itself rather than to its objects. Beneath this artificial stability is a continuous succession of states, each announcing what follows and containing what precedes, uncountable until attended to as passed.[37] Conscious states, in enduring, belong to a perpetual becoming, succeeding one another without mutual externality. They express the whole person, each state virtually containing the person's past and present. By carefully attending to pure experience we can reduce the distortions due to practical concerns, intuitively entering more intimately into the interpenetrating qualitatively differentiated whole of *durée*. The intuitive method separates out pure differences of kind, virtual tendencies operative in experience but only found there combined with other tendencies. Such differences in kind are pure a priori conditions of actual experiences.

Within every concrete perception Bergson distinguishes pure perception, memory-images, and pure memory. Pure perception prolongs sensation into action. For only the present is of practical interest. As present, pure perception is a localized source of activity in the body, differing in kind from pure memory in which the real past exists, inextensive and ineffective. Pure perception only exists in the ideal present as the limit separating past from future, while concrete consciousness actually lives in the active present. This living present occupies a duration including the immediate past and immediate future. Consequently, the living present cannot be separated from memory. What neither acts nor concerns action does not cease to exist, but belongs to the past, virtual and unconscious in pure memory. Pure memory remains unattached to the present when useless, but, when needed for action, can be actualized in *sui generis* acts of recollection. We actualize memory by detaching ourselves from the present and leaping into the past. In recollection, we may become conscious of the past through the passing of the virtual past into the actual present in memory-images, nascent perceptions, in which pure memory becomes more distinct and adopts an orientation appropriate to action. Yet as the virtual past expands into an actually present image it must retain something of its virtual state to be distinguishable as memory rather than sensation.[38]

Memory-images coincide with pure memory in a region of indeterminacy where two divergent axes meet, one actual, the other virtual. In the first, memory-images connect practical needs with actions in a sensory-motor connection in which the body's reactions adapt to the practical requirements of the present situation. In this connection, habit, a set of mechanisms of adaptation, behaves as a form of memory. Habit organizes sensory-motor systems through an interval where memory-images delay the passage from perception to action. Recollection fills this interval, selecting what is useful from the past and making qualities appear by interpolating the past in the present. Pure memory preserves the past with all our states marked in their order of occurrence. Pure sensory-motor connections exist only in the instantaneous, but ideal, present, while pure memory moves entirely in the past. In the interval, the virtual mixes with the actual, producing an impure and indeterminate region. The two coalesce in living perception, in which all that is perceived is immediately past.[39]

There is, then, no instantaneous point dividing the past from the present. The living present is a continuous becoming in which past and present commingle. Since it is all that is actually given to consciousness, only this living present seems to be real. Yet, because of their imminent promise or threat, we readily concede the reality of things actually lying beyond consciousness. This practical utility unfolds the future as impending actions, and their distance in space as a measure of the proximate need for such actions. Regular causal and logical connections between what follows and precedes it extend actuality beyond the living present. Consequently, the contents of actual perception belong to a wider experience, which is as yet unperceived but seems already to exist.

More difficult to acknowledge is the reality of the past. The same practical utility that opens the living present on to the future closes it off from the past. For, in itself, the past has no efficacy. It becomes efficacious only if it is taken up by recollection and mixed with present perception. Practical interests dispose perception to look at what is unfolding, and hence only at the immediate past and those aspects of the past that answer to present needs. Memory, therefore, appears in consciousness discontinuously, concealing the real continuity of the past. Yet, like unperceived actual objects, the past coexists with and beyond the living present. But not everything real is actual. Although it is not conscious, pure memory has as much claim to reality as unperceived objects. The past exists virtually, even though it is no longer perceived.[40]

The whole of the past conditions the present, even though only fragments of it appear in consciousness. The prolongation of the past into the present, the interpolation of pure memory, is the condition of the passing of the present, making the past and present commingle in the living present. Otherwise, conscious life would shrink to an instant. Memory conserves the past in the present; so the passing present always coalesces with the immediate past – the latter has not disappeared when the former arises.[41] Since the past cannot be reconstituted solely from elements of the present, such as present traces deposited in the sensory-motor systems, we must, through recollection, leap into the past itself. The living present continuously divides along these two directions, that of actuality directed toward the future, that of virtuality directed toward the past. While the past conditions consciousness, it is preserved only in itself. Denial that the past survives results from confusing being with being-present. The psychological component of recollections consists in their being useful for action, their being present. However, in addition to their being present, recollection also invokes pure memory, an inactive, non-psychological being – a purely virtual ontology of the past in itself.[42]

The past becomes present in psychological memory only through its actualization. The past differs in kind from the present, not merely in degree. While pure memory conditions the passing of the present, sensory-motor habits provide a means for pure memories to become embodied in the present. Present perception abstracts resemblances in which recollection injects distinctions. When pure memory furnishes unconscious memories they become present and active in the sensory-motor connection. Sensory-motor habits use past experience in acting, but only pure memory can provide its image. Were we to live only in action we would become creatures of impulse, habitually reacting to present situations solely on the basis of their resemblances to earlier situations. Were we to live only in memory we would be dreamers, dwelling entirely in the past, recalling all its particular differences at once, but never acting. However, we never live exclusively in either.[43]

Instead, past and present coexist, coalescing in the living present. It is insufficient to distinguish the past from the present merely by indexing a time and locating that point among before and after relations. That would conceive the past as caught between the present it was and the present of which it is the past, as if all there were to the past were having been present or occurring before the present whose past it is. That would treat the past merely as another present differing only in degree from the indexed present. Bergson continually warns against this propensity to recompose the past from a series of presents. The error arises because the past can be actualized only by adapting to the practical needs of the present. But the past is virtual; it is the pure condition necessary for the present to pass. For the present to pass it must do so while it is present. Instead of a succession of presents, a new present can arise only at the same time as the old present passes. The present continually passes, coexisting with a past that includes all presents.[44]

Bergson pictures this coexistence of past and present in the form of a cone, SAB, meeting, at its vertex, S, a plane, P (see Figure 8). The cone represents the totality of the past abiding and coexisting with the present. S depicts the passing present, where the cone of the past joins plane, P, depicting the representation of the universe from the point of view at S. The present perception of the sensory-motor connection centers at S, so the image of the body belonging to P and memory-images recalled from the past concentrate at S. S is the point at which actions from P impinge on the image of the body from which actions return to P. The base, AB, depicts the totality of recollections. S and P advance continuously, while the base, AB, remains immobile. Hence habit governs the sensory-motor connection centered at S on P, while pure memory coexists with it, offering recollections of the past capable of guiding action.[45]

At S perceptions prolong themselves into appropriate reactions. Perception extracts what is useful from its situation and stores it in habit for future use. The closer its focus to pure perception, the more habit assembles repeating elements. Recollections, on the other hand, interpolate the past in the sensory-motor connection. Through this interpolation, recollection combines the multiplicity of past events with perception, which selects those necessary for action. But if dissociated from the requirements of action at S, recollection need not dwell on one part of the past more than another. Hence, recollection expands into the totality of the past, AB, and any memory may accompany the present situation. The more consciousness attaches itself to the actual present, responding

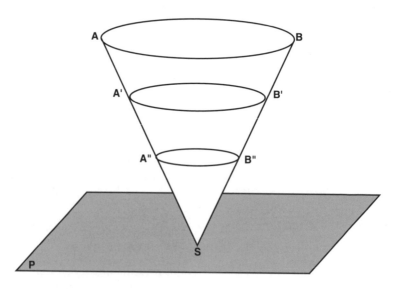

Figure 8: The cone, SAB, of the past coexisting with the plane of the representation of the universe, P

Source: Bergson 1988: 162

to stimuli via motor reactions, the more it concentrates at S. The more it detaches itself from action, the more it diffuses over a multiplicity of well-defined images in the past, AB. (Virtual multiplicities are thoroughly differentiated, coexisting without merging, like the superposition of wave forms.[46])

Consciousness continually oscillates between these extremes, which are never wholly attained. This oscillation sets up a circuit from S to AB as consciousness expands and the past increasingly presses to insert itself in the present, and from AB to S as consciousness contracts to meet the fundamental needs of life. Between S and AB there may be indefinitely many virtual repetitions of the totality of the past, $A'B'$, A''; B'', etc., at different levels of expansion-contraction (see Figure 8). Consciousness narrows or enlarges its content in this double current, but, no matter how contracted the memory, the whole of the past coexists with the present. Recollection leaps into the past in general, then into a certain level containing the whole past in a more or less contracted state. In every level of the past all memories gather around certain dominant memories, acting like shining points. Dominant memories multiply the more memory expands. The greater the number of these points that are prominent, the less contracted the level of the past. Recollection leaps to a level sufficiently expanded for it to search across the level from its shining points to locate the memories it needs. The pure past becomes actualized psychologically only if recollection finds its appropriate conditions, useful dominant memories, in a region of the past into which it leaps.[47]

DURÉE AND ONTOLOGY

Actualization distributes a continuous creation of novelties. While it is similar to emanation in Neoplatonism, possibilities of being do not pre-exist their actuality, nor does being degrade as it appears in time. The virtual is a temporal becoming actualizing itself without loss. For Bergson, *durée* has ontological primacy over form or matter. *Durée* consists in the virtual becoming actual. Interpenetrating differences in the virtual whole become actualized by creating divergent lines corresponding to differences in kind, and embodying outstanding points from particular levels.[48] Differences that cannot be dissociated in the virtual whole become separated in their actualization. While the virtual whole unfolds along these lines, its actualizations neither resemble it, nor can be recomposed back into it. There is no resemblance, since actualization creates resemblances and possibilities themselves. Actual things may resemble one another without being produced by similar lines of actualization.[49]

Virtual tendencies create their own lines of actualization by movements of translation and rotation. In translation, a whole level of the past coalesces with the present, without dividing; it blends reciprocally with present experience. This coalescence translates a level of the ontological past into a psychological image, but does not pass through other levels of the cone. While a particular level of the past translates into a memory-image, coalescing with the present, recollection turns toward the present, offering its most useful facet to be put into effect. This rotation orients recollection images, expanding useful distinctions from the past into the present. Present perception acts reciprocally on these images, harmonizing translation and rotation. Moreover, actualization requires the incorporation of memory. Useful motions and attitudes of the body adapt these images to its habits, filtering out useless memories. So both past and present participate in actualizations, in which the past becomes embodied and displaced by a new present.[50]

While consciousness unfolds in this way, actualization is not subjective – things must unfold in an ontological *durée*. Time acts, it retards and elaborates; it flows without any *thing* flowing. It is the indetermination of things, actualizing differences and assuring the survival of the past in coexisting levels of expansion and contraction. Through intuition of our own *durée* we may rise to acknowledge other *durées*. If we mix sugar in water we must wait for it to dissolve. Bergson maintains that the necessity of this waiting links conscious *durée* with that of the universe as a whole. *Durée* in consciousness and *durée* in the universe are aspects of the same multiplicity. Common sense, Bergson notes, supports the hypothesis of one universal impersonal time. It supposes, first, that all consciousnesses live the same rhythm of *durée*. Second, were each pair of consciousnesses to share at least one experience of something external, they could form a chain that would cover the whole universe. Finally, since external things participate in the same *durée* as a consciousness that perceives them, the consciousnesses can be eliminated and the whole universe participate in one universal impersonal time.[51] Bergson accepts this conclusion, but turns to the special theory of relativity to confirm it, and to show that physical theory accords with his account of *durée*.

The special theory of relativity seems to contravene the commonsense notion that we can treat large and small distances similarly, and treat simultaneity of near and distant events likewise. The Lorentz transformations show that the time elapsed will be greater in a coordinate system X than the time elapsed in a coordinate system X' moving relative to X.[52] In 1911, the physicist Langevin suggested that this implies that Paul, traveling in system X', will age less than Peter in system X. For Paul's time and metabolism will have slowed compared to Peter's. Yet, since it is conventional which system is at rest and which in motion, holding X' to be at rest implies that the time elapsed will be greater in X' than in X and that it will be Paul who ages more. Bergson asserts that the special theory does not support this hypothesis of different rates of time.[53] He agrees that the mathematics of the Lorentz transformations will give different measurements of the time elapsed in X and in X', but argues that careful interpretation of these measurements shows that they do not support the hypothesis.

Bergson argues that only a real time can be perceived; a time that is imperceptible would be imaginary, merely symbolic. Moreover, for time to be a genuine succession something like a memory or consciousness must unify before and after. Physicists measure time by substituting a line for time given in perception. But it is justifiable to call this line time only if its juxtapositions can be converted into succession. Bergson insists that the simultaneity of two instants used in physics does not belong to real time, but is a symbolic abstraction. Relativity physics measures the abstract, merely quantitative simultaneity of two clock readings according to a convention for determining under which circumstances such readings should be called simultaneous. This abstract time is not lived by anyone. It is not the real time lived by either Peter or Paul; nor is it even the time either imagines the other really living. But real time must be lived or conceived as capable of being lived.

Langevin's suggestion only arises through a confusion between real time and that of the mathematical symbolism of the theory. This symbolism can be interpreted either, with Lorentz, as representing real physical changes or, with Einstein, as providing a convention for measuring the simultaneity of distant events. Bergson rejects the former as inconsistent with experience. Experience gives only real *durée*, i.e. qualitative successions that symbolic abstraction treats as juxtapositions, and with it the contemporaneity of fluxes. Experience of the contemporaneity of fluxes is necessary to measure elapsed times, because observing that a phenomenon occurs at the same time as a clock movement requires that these two fluxes be encompassed by the observer's own flux. These fluxes can be lived only in the perspective of a single time. For different fluxes to be contemporary we must be able to attend to them at once as united in one whole and differentiated into multiple fluxes. Hence, because it rests on experience of fluxes, physics cannot reject *durée*.[54]

For two fluxes to be unified, a third must encompass them. By encompassing other fluxes, *durée* unifies them in contemporaneity or succession. Consciousness exemplifies this unification. Moreover, through reflection, it encompasses itself without limit. In this way, consciousness discloses but is not identical to *durée*, to which it gives intuitive access. *Durée* does not reduce to one or many; it is pure multiplicity unfolding in its actualization. This unfolding consists in the creation of lines of actualization – a single time, but a virtual multiplicity, expanding into an infinity of fluxes along lines of actualization. The universe acts like a single memory, a virtual coexistence opening out onto actuality, always quality and quantity commingling in perpetual flux. In its actualization *durée* spreads its moments over an increasingly quantitative continuum, dissipating the tension of their interpenetration so that one must disappear for another to appear.

The theory of relativity pictures the whole of time laid out in Riemannian space. But even this does not entirely spatialize time. For the formula for space–time separations between two events marks a qualitative difference between spatial and temporal components by the sign of the temporal component.[55] In its analysis the theory of relativity attempts to represent time fully actualized. It analyzes qualitative into quantitative multiplicity, treating the universe as entirely actual, already complete. By restoring the unfolding of the virtual in it actualization, this near spatialization becomes compatible with Bergsonian *durée*. *Durée* is an enduring becoming – the virtual continuously actualizing itself, but never entirely actualized.

CHAPTER XIV

Becoming-time

Duration is construction.

Paul Valéry

Recently, Gilles Deleuze (1925–1995) has refined and built upon themes from Bergson's approach to time. Although Bergson is often considered a maverick among philosophers, Deleuze shows that he belongs to a distaff tradition in philosophy.[1] Deleuze produces a genealogy of this distaff tradition in studies delineating thematic elements that connect various thinkers, each expressing motifs out of joint with their philosophical contemporaries, yet pursuing lines of thought that leap from one distaff thinker to another. In part, Deleuze incorporates, consolidates, and advances this genealogical enterprise through his approach to the transcendental problems associated with time. In this respect, Deleuze extends insights from the distaff tradition, transposing the transcendental role of time into a type of empiricism that transmutes the conception of the transcendental field from that of Kant or phenomenology. This transmutation will suggest some complex correlations between expositions of time in the three twentieth-century traditions.

THE FIRST AND SECOND SYNTHESES OF TIME

For Deleuze, the problem of time is not whether time consists of an A-series or a B-series. Present and past each contain B-relations, but consist themselves of neither an A-series nor a B-series. Although Deleuze writes of past, present, and future, these terms primarily refer to syntheses, not moments or even designations of moments.

Husserl recognized the transcendental function of passive syntheses, but treated the temporal flow as intentionally constituted. For Husserl, the passive syntheses consisted of identifying and distinguishing syntheses; such syntheses are neutral regarding all positing,[2] but the extent to which they are also generative is obscure. Since absolute consciousness constituted the temporal flow in experience through syntheses ordering temporal phases and individuating their contents, time was grounded in subjectivity. Deleuze renounces grounding time in subjectivity, in any of its guises, transcendental or immanent. He denies that the transcendental field has the form of the subject, but acknowledges that any empiricism must recognize something performing the synthetic function of a transcendental field, something both neutral and generative. Deleuze's empiricism asks how the abstract, including all totalities and unities, can be explained by states of things that are always multiplicities. This empiricism seeks to discover the conditions under which something new is produced.[3] Hence it can ultimately appeal in explaining the constitution of time neither to the subject's unity, nor to limitations, nor to orderings the subject imposes. Instead, the problem of time concerns the generation of time as the Event in which something new can be produced, prior to all subjectivity and in which the subject is produced as well.

The first synthesis of time produces the living present, by connecting successive instants as past and future. Time cannot be constituted solely as a succession of instants. Were that the case, time would begin to arise at each instant only to fail to continue one instant into another. Moreover, instants are abstract. Each instant marks the arising of a singular element that must be connected with subsequent such elements. The connective synthesis is a contractile power that binds independent elements together into a single time, retaining one element when another appears. It constitutes a past by retaining and contracting preceding instants, and a future because this same contraction opens onto and can continue appropriating succeeding elements. The order from past to future is the order of action, that from particular to general, from actual to possible. Hence this continuous synthesis produces a living present containing particular past instants, yet general insofar as it is open to the future. However, elements never attain their status as past and present independently, but only as a result of the synthesis that produces their connection. Past and future are dimensions of the living present that connects these elements.

The first synthesis of time functions to fuse successive elements by extracting the difference between past and future, hence producing the repetition of seemingly identical elements. There could be no repetition of completely identical instances, since, were that so, nothing would distinguish one from another. There must be a difference between repetitions that distinguishes them, but by extracting that difference the synthesis binds them together into a continuous homogeneous series. This despite the necessity that one instant can only continue into another through a differential displacement generated by a dissymmetrical difference internal to each element.[4]

The connective synthesis is a passive synthesis of time, prior to reflection and memory, functioning to bind elements on various levels. Below the level of sensation, organic syntheses bind organic elements, fusing, for instance, contraction and dilation into a past and future of the actions of muscles, cells, or nerves. In other instances, other organic syntheses fuse organisms into a past of cellular heredity and a future of organic need, while sensory syntheses bind elementary excitations into sensed qualities that further syntheses associate successively on yet another level. Although each connection produces a living present, the connective synthesis originates from neither a reflective subject nor an activity of the mind. On the contrary, fusing repeated elements into a living present makes actions and active subjects possible, by giving them a foundation in connections among successive presents within the living present.[5]

The first synthesis distinguishes past from future and connects them; it provides the foundation for, but does not explain, the living present. For the living present, constituted by the first synthesis, also passes. This passage is more than a successive order of former presents connected to the present present. For the present to pass, it cannot pass by itself as if it were coextensive with time. The living present could be coextensive with time only as an infinite succession of instants, but instants could extend infinitely only insofar as they were abstract. The elements contracted by the first synthesis form a limited present, variable in extent. These contractions both constitute the living present, and must themselves pass in it. Within the living present an active synthesis, consisting of phenomenological consciousness and the dimensions of representation, is founded on the passive synthesis and the connected order it produces. For the first synthesis makes every present possible. The active synthesis constitutes the present and former presents intratemporally. This requires that the living present contain a dimension other than that of the line of succession, a dimension in which each present can reflect on itself and represent former presents; this dimension evokes intentionality. The living present retains particular former presents; to pass, however, it must also reproduce and represent present and former presents as belonging, not to the particular past contracted and retained in the living present, but to the past in general. This evokes Husserl's distinction between retention and reproduction, although it grounds intentionality rather than being grounded in it.[6]

The first synthesis of time, then, occurs in time. Since it does, there must be another time, the past in general, in which it occurs. The past in general differs in kind from the living present. In the living present the present is not an instant that will be a former present. It is that which represents both itself and former presents as passing. In representing this passage, an active synthesis embeds presents in themselves, referring to them from a dimension perpendicular to the line produced by the first synthesis of time. So, while the active synthesis, i.e. consciousness, requires the first synthesis as its condition, it must also be made possible by, and require as its condition, a second synthesis of time, constituting a pure a priori past.[7] This past is not a former present, but that dimension from which it is possible to represent the present and former presents. It constitutes a general region in which particular presents preserve themselves so that it is possible to focus on and represent them in the present present. Consequently, the pure past is a necessary condition for the possibility of any representation that constitutes the living present, and it cannot be derived from the first synthesis alone, or from the present, or from any representation indexing moments in the living present. This pure past, then, must be constituted by a second passive synthesis conditioning the preservation and representation of present and former presents.

The second passive synthesis constitutes a pure a priori past in time. It is a transcendental synthesis, operating through constitutive paradoxes that show why the pure past must have its own being contemporary and coexistent with the present even though it was never present.[8] The present cannot pass merely because a new present appears. For a present to pass it must be constituted as past, and it cannot be constituted as past unless it were so constituted when it was present. For everything past has already been constituted. The past, then, must be contemporaneous with the present it was. This explains why every present must pass; every present is already past when it is present. This paradox leads to another, that the past coexists with each new present with respect to which it is past. For, if every present is already past when it is present, then the past is not in any one present and coexists with each new present.[9] These paradoxes differ fundamentally from McTaggart's; for they do not treat past and present as similar determinations in a single series, and likewise with respect to the future, which enters into these constitutive paradoxes only through a third synthesis of time. Moreover, these paradoxes function as syntheses, generating the activity of consciousness and the conditions of representation.

The pure past is neither a former present nor now present; it does not stand in before and after relations within the living present. The pure past was never one of the successive presents in the living present. Instead, it is presupposed by every member of this succession and is a general condition calling up each new present and grounding its passage. Consequently, the pure past is a synthesis of the whole of time rather than part of a series of times. It neither arises nor passes, but contains all particular times; it is the whole of time in itself, outside the living present, the whole of the past coexisting with each present.[10] From the perspective of the first synthesis, only the living present exists and any past outside its scope is no more. From the perspective of the second synthesis, the living present is outside itself a priori, becoming without having come to be. The past already exists in itself, as a condition prior to but subsisting in former presents and persisting in each new present.

The living present presupposes the past as already real, preserving itself as something without actual existence. Accordingly, the mode of being of the pure past is virtual; its being is a condition of every present. As presupposed to the living present without belonging to it, the virtual past has a reality of its own, an ideal but non-psychological being. Characterizing this ideal mode of being, Deleuze cites Proust; the virtual is "real without being actual, ideal without being abstract."[11] Since it is virtual, the past comprises the whole of time, but this whole differs essentially from any set of elements or instants. However diverse its members, a set is always a relatively closed totality. Nevertheless, each set may be linked to a larger set and such linkages may continue virtually to infinity. The whole is not a set at all; it is what keeps each set open on to more inclusive sets; it guarantees the passage, extending one set into another.[12] This explains how the virtual past makes the present pass. Since it is the whole of the past that coexists contemporaneously with every present, it opens each set of presents on to a new present.

Moreover, since the whole of the past is open, continually differing from itself, it guarantees the irreversibility of time.

The first and second passive syntheses refine Bergson's image of the cone and plane,[13] specifying their operations and the complex nature of the present. The first synthesis contracts successive independent elements or instants, so that the present consists of their most condensed state. This synthesis constitutes a series of empirical presents by founding actual relations of succession, simultaneity, contiguity, resemblance, and causation between them. The second synthesis constitutes the whole of the past coexisting with each present by going beyond the empirical present to the a priori past. All differences coexist in the virtual past in varying levels of compression and relaxation. While the first synthesis extracts the difference from successive elements, the second synthesis inserts a differential level of the whole in the present. Relative to the first synthesis the present links successive elements and the past spreads out into a single continuous series, while relative to the second synthesis the present consists of the most compressed degree of the multiple coexisting levels of the past.

THE THIRD SYNTHESIS OF TIME

Since the present includes both the contraction of successive presents in the living present and the most compressed level of the past, it is double, at once actual and virtual, present and past, material and ideal. This doubled present emerges in what Deleuze calls a crystal of time. The virtual and actual form two distinct sides of this crystal. Actual and virtual continuously exchange and displace one another within this crystal, resulting in indiscernibility between the crystal's two sides. Crystal-images exist in which such discernment is impossible, an objective illusion arising out of an objective zone of indetermination. Determining whether we are seeing a thing or its image in a mirror serves as a paradigm. Were everything that does not appear in the mirror inaccessible we would not be able to discern the thing from its image. The image is virtual with respect to the actual thing, but the thing seen only in the mirror is virtual with respect to the actual image. Either the actual appears and the virtual gets referred elsewhere or the reverse. Without access to conditions outside the image, which is clear and which obscured always remains in doubt.

Whether the present is actual or virtual can only be discerned from outside the crystal, outside the present, only by reference entirely to the whole of the past or entirely to former presents. But, the present is double. Every present gets referred both to the past in general and to former presents; the crystal both extends to former presents and expands to more relaxed levels of the past. So, within the crystal, the indiscernibility of virtual and actual persists. The virtual past corresponds to its particular present like a reflection in a mirror, but the actual present also corresponds to its virtual past as if it were the reflection. Although each taking the other's role does not vitiate the distinction between virtual and actual, it makes ascription of only one to the present impossible. Within the objective structure of the crystal, they remain both distinct and indiscernible.[14]

Since the present contains both the actual present and the virtual past there must be an operation by which the present splits apart, making the actual present pass while preserving the virtual past in itself. This operation forms the third synthesis of time, occurring within the crystal, its effects perceptible within the crystal-image. This splitting is the most fundamental synthesis of time, for it differentiates the other two. The third synthesis constitutes the past in the same time as its present is displaced by its successor. Time splits into two heterogeneous dissymmetrical emissions, one toward the future, making the present pass, and another toward the past, coexisting wholly with the present it was. The third synthesis, then, generates the other two. Essentially, time consists in this splitting, in the essential disequilibrium between actual and virtual, their displacement and the impossibility of their equivalence.

With this third synthesis, Deleuze grounds the first two Bergsonian syntheses in a Nietzschean eternal return, inducing an empiricist resolution of the transcendental problem. Our experience of time represented within the active synthesis is of a passing succession of presents. The first and second passive syntheses of time give its conditions. The connective synthesis joins these presents together in a living present, and the conjunctive synthesis of coexisting past levels makes the present pass, giving the active synthesis a dimension from which it can represent this passage. But how can the past make the present pass if it has no efficacy? Although itself a passive synthesis, since its operation ultimately grounds the activity of consciousness generated in the active synthesis, the third synthesis is disjunctive and productive. It splits the virtual from the actual, and in doing so keeps the past open, making the present pass in the production of something new. Moreover, by doing so in the present itself it shows how the ideal reality of the virtual past can belong to experience. While the first synthesis constitutes the present and the second the past, the third passive synthesis constitutes the future, producing something new by splitting virtual from actual and making the present pass. The third synthesis evokes a Nietzschean eternal return, in that ever new becomings arise in each moment.[15] Present moments consist of tensions continuously dividing into actual presents and the virtual past. Within each present something new arises, a pulsation in the depths of matter becoming spread out in past and future time.

Becomings arising from the present into the past and future evoke the two Stoic readings of time.[16] The first reads time as Chronos, corporeal time that is wholly present. In this sense, we say that only the present exists. In the material universe of bodies and states of affairs, bodies mix and permeate one another in a unity of causes. Everything that is happening happens in the living present. In the active synthesis (and as pragmatic analyses of time show) action arises in the living present, providing the direction from action to passion and the extension in which actions and passions can be measured, giving things definite limits. This extension results from the first synthesis of time, which links presents together so that past times consist only of former presents. To the degree that there is a unity of all causes, a unity of principles of action and passion, Chronos extends this living present to the whole of the universe. Yet Chronos can extend the living present only to the extent that it provides its own measure for the unity of causes. Chronos can provide this standard only if relations between causes (actions and passions) in the mixture of bodies limit the present. Such a limitation requires that each present or cosmic present return exactly the same. For it could only differ were it subject to additional causes belonging to former presents or presents yet to come, contradicting its claim to incorporate all causes.

Although everything exists in it at once, Chronos is not eternity but a dynamic and multiple causality, a living present. Since this living present is not already laid out it may not encompass an absolute present. The unity of causes brings about surface effects, and Chronos limits their actualization. Insofar as it provides a measure, Chronos forms a cycle. Insofar as something new arises, the novelty creates a break in this cycle and becomings arise, subverting the unity and measure of the existing present. As a result of this break, becomings appear on the surface as events happening to bodies. For events to rise to the surface from the depths of bodies becomings must change their nature, becoming pure ideal events differing in kind from their causes in bodies and states of affairs. Rising to the surface, pure becomings gain a new orientation, laterally past and future, but never present. Past and future do not exclude one another, but emerge together between times.[17]

Occurring on the surface, becomings constitute pure events, continuously dividing into past and future. Fundamentally, time consists of this pure continuous operation, splitting virtual from actual and opening at once on to past and future, an operation comparable to an ideal game. Ordinary games form part of human activity, having established rules that obligate players. These rules distribute the possibilities of gain and loss according to really and numerically distinct plays (such distributions parody practical utility or morality). Chance may operate here only at certain points. Ordinary games form closed sets of fixed possibilities susceptible to analytic procedures. Alternatively, an ideal game constitutes events that ramify chance at every point, prior to any human activity. While the game is rule-guided, the rules do not pre-exist play; each move or play invents its own rules. Hence plays are neither really nor numerically distinct; they do not distribute gain and loss. The ideal game forms an open whole, an eternal return, through which actualities and possibilities become constituted by a disjunctive synthesis of problems and questions.[18]

The second reading grasps time as Aion, as an incorporeal time of ideal events. Aion contains only effects, incorporeal becomings appearing on the surface of bodies as a result of actions and passions. Since they lack all efficacy, such pure effects unfold as ideal events, existing neither as states of affairs nor as properties of things, neither as agents nor as patients. Events inhere in things; they subsist as impassive idealities, neither acting nor being acted on. Pure events are infinitely divisible, always both past and future but never actually present; they move in both directions at once. Since these pure becomings neither exist in the living present nor move in a single direction, they transcend material limits and have no measure.[19] Events operate as plays in an ideal game. Endlessly ramified, dividing into past and future, each event eludes the present, occurring between times in a time that cannot be measured.[20]

Although they result as effects from material causes, events resonate independently in connection, conjunction, and disjunction with one another, allying in Aion, the Event of all events. Aion is the empty form of time, the a priori unchanging form of all change. Aion, unfolding as such, plays the ideal game, endlessly ramifying chance. Unlimited, it is that which keeps sets of possibilities and actualities continuously open on to more extensive sets. As such its geometry is not that of an uninterrupted line from past to future. Past and future constitute a priori dimensions within Aion, not empirical specifications of time.[21] Aion constantly decomposes, dividing ad infinitum, continuously distancing past from future. Its geometry is fractal, unfolding in a dimension greater than zero but less than one. Aion comprises the form in which events occur and communicate with one another, the locus of incorporeal effects. Aion consists in a caesura, a crack between times, the Event in which events arise and communicate in a time less than the least conceivable time.[22]

Aion occupies less that the least conceivable time, because it delineates the displacement between one present and another, eluding measurement in the living present. But Aion also constitutes a singular event in a time greater than the greatest conceivable time. For it supplies the a priori form of events, encompassing all events within a singular form, exceeding measurement by the limited present. Aion encompasses all events successively with respect to one another, because it continuously distances past and future. Furthermore, Aion also affords a singular point with respect to which all events are simultaneous, since it encompasses all events in the crack between times. Events occur in Aion, but since they are singular, hence unlimited, dividing infinitely, they do not compose or fill it up it.[23] Each event is a singular point, a unique play in the ideal game, but also a pure multiplicity, an endlessly ramifying distribution of singularities. Ideal events are sets of singularities that unify themselves in a continuously changing distribution. Aion, the Event, is the singularity in which all events communicate, the empty form of time.

In this respect, Aion consists of a singular point continuously displacing itself throughout the series of events, coordinating and endlessly ramifying the series. This singular point tracing the straight line establishes its fractal form. Splitting Aion from Chronos, this point occupies the point of indiscernibility in the crystal of time. It is aleatory, occupying the limit between bodies and events, injecting pure chance between times. This singularity consists in a paradoxical element, an essentially partial object that unifies series of events by continuously displacing itself along and between those series. Since it has two sides, two dissymmetrical parts, one actual and material, the other virtual and ideal, the paradoxical element is partial, split apart, and incomplete. Moreover, it is both partial and paradoxical because it is always in disequilibrium with respect to itself, continually displacing its own identity. By continuously displacing itself, the paradoxical element generates endless disjunctions.[24] It acts as a differenciator,[25] producing an eternal return of difference, the third synthesis of time. The paradoxical element can cause the present to pass, to be displaced by another present, because it is partial. The paradoxical element opens each present onto the future by exceeding the actual present through the production of its virtual double. The third synthesis of time, then, generates the other two. Since the paradoxical element is both virtual and actual it constitutes the simultaneity and coexistence of past and present, and its endless ramification splits past from future, creating lines of actualization. Its continuous displacement connects each present with its virtual successor, assuring that the present passes because its difference from the former present eternally returns.

THE TRANSCENDENTAL PROBLEM

Since it comprises the most fundamental synthesis of time, the paradoxical element constitutes a boundary condition on phenomena, their dark precursor and transcendental condition. However, Deleuze rejects both Kantian and phenomenological presentations of the transcendental problem and its resolution. He repudiates conceptions of the transcendental field as providing conditions of possibility, preferring the discovery of pure tendencies, virtual conditions for actualities. This more restrictive variant adheres closely to Deleuze's empiricism because the conditions of experience, virtualities rather than possibilities, never apply to anything beyond what they condition.[26] Everything is double, virtual and actual, but not by virtue of unity or resemblance. The transcendental field must generate unities, resemblances, and identities. The transcendental field does not have the already unified form of a consciousness or subject, whether as act of thinking, pure concepts of understanding, or nexus of intentionalities. All of these must be produced from pure multiplicity. The problem, then, is not to recover the universal or to ground the manifold in a prior unity, but to find the conditions under which something new is produced.

This transcendental empiricism harks back to Duns Scotus' ontological proposition: being is univocal.[27] This does not mean that there is only one being, but that being speaks with a single voice, that it applies in one sense to the divergent multiplicity of

beings. Being occurs as a singularity, the Event, from which everything diverges, ramifies, and in which other singular points resonate.
This singular Event occurs when a paradoxical element generates
surfaces between topological planes, between bodies-causes and
events-effects. Singular points exist on the surfaces of bodies but
also subsist there as incorporeal events-effects. Since they are incorporeal, events should not be confused with their actualizations in
states of affairs. While the unity of causes exists in the depths of
bodies, the Event generates the coexistence and communication of
effects through a quasi-causal linkage along the surface. This quasi-
causal linking arises when a paradoxical element ramifies series of
events, distributing, articulating, and coordinating them in corresponding topological planes, making them coexist and
communicate, producing an autonomous and intelligible organization on the surface.[28]

Although a quasi-cause generates an intelligible organization, it
is itself a paradoxical element that includes nonsense co-present
with sense. The Event is sense arising out of nonsense when a paradoxical element generates the autonomy of the event's organization
on the surface. The being of this organization is univocal in relation
to both what occurs and what propositions say. Organization of the
surface occurs, regarding propositions and things, on various topological planes. Outside the proposition the paradoxical element
quasi-causes a univocal articulation of bodies and language. As
phenomenology contends, both things and propositions are meaningful, with corresponding articulations. Since sense inheres in
propositions and is also attributed to states of affairs, it occupies
the boundary between propositions and things. Circulating on the
surface between things and propositions, the paradoxical element
generates an articulation of bodies into objects and properties as it
engenders an articulation of propositions into subjects and predicates. Similarly, events, at once past and future, happen to things
but inhere in language, as expressed by infinitive verbs prior to any
determination of tense.[29] Preceding surface articulations, the unin-
flected infinitive expresses univocity transmitted from being to
language. Lastly, within propositions, the paradoxical element
passes between denotations of things and expressions of sense.[30]

Deleuze analyzes three dimensions of the proposition: denotation, manifestation, and signification. Denotation relates words to
things or states of affairs, subjecting propositions to values of truth
and falsity. Manifestation relates propositions to the beliefs and
desires of a speaker, subjecting them to values of veracity and illu-

sion. Signification relates words to general concepts, which syntactically connect propositions serving as premises or conclusions of demonstrations, subjecting words to the value of possible truth or falsity as opposed to nonsense. However, none of these dimensions of the proposition contains sense, which establishes connections between words and things. Were denotation to provide the condition for propositions to attach to states of affairs it would lead to paradox, since it would presuppose the very correspondence between words and things it needs to establish. Manifestation is primary in speech, where the "I" functions as a pivotal point presupposed by all denotation. However, the speaker cannot manifest beliefs or desires except through the use of established concepts. While the conceptual connections of signification provide truth conditions, signification cannot establish the connection between words and things because that connection requires denotation, that the premises be true for a true conclusion to follow. For, actual truth and falsity are not determined by their conditions. More important, the truth of the proposition that the conclusion follows is itself an implication that must be demonstrated from some premises, leading to an infinite regress.[31]

Signification supplies truth conditions only as possibilities, where the proposition rises from conditioned to its relative conditions. But, rather than this regress of conceptual possibilities, language and the conditions of truth ought to be generated all at once by something unconditioned. In addition to the de facto circular relation between denotation, manifestation, and signification, Deleuze proposes that the Event answers the question, *de jure*, how language is possible. The Event conditions the proposition, distinguishing its three dimensions, by generating the incorporeal surface between ideational attributes of things, e.g. the intentional correlates of perceptual acts, and their expression in propositions, and that between denotations of things and expressions of sense. The Event is the expression of sense, making language possible all at once. It attributes sense to states of affairs and expresses it in propositions. Sense is ideal being rather than a quality of things, yet it has an objectivity separate from the proposition, inherent in things. It is the pure Event that organizes language and the intelligibility of things around its continuously decentered singularity, along the incorporeal line of Aion, in the crack between times.[32]

Singularities, in the Event, condition all phenomena, their unities, and identities. Earlier thought erected the transcendental field upon one of the dimensions of the proposition; Kant elevated signification to transcendental stature, phenomenology elevated denotation, and Lockean empiricism elevated manifestation. Deleuze objects that erecting the transcendental field on resemblance to the empirical field it engenders is circular. The transcendental problem cannot be solved by appealing to a dimension of the proposition to ground its intelligibility or that of things. Instead, the transcendental problem asks how something engenders the proposition in all its dimensions.

Objectively undetermined, singularities do not belong to the proposition, but to the problematic. The problematic is neither a subjective category of knowledge nor an indication of deficient method. Instead, it is both an objective kind of being and a category of knowledge presiding over the genesis of solutions. It functions as the object of unconditioned Ideas to which no empirical proposition is ever adequate,[33] and forms the horizon of all that appears. The transcendental field, then, must be a pre-individual, impersonal, aconceptual continuum of singularities, a pure multiplicity, a zone of objective indetermination, rather than the form of the identity of an object in general.[34] This field emits singularities, distributes differences, and establishes resonances between them, auto-unifying the Event. The transcendental field actualizes the virtual, generating series and unities, including the unity of consciousness. Actualizing the virtual, the field organizes an intelligible world, forming a topological surface expanding at the edges like a crystal.

This transcendental field must be neutral, suspending as problematic every aspect of intelligibility, whether in propositions or things. Singularities-events occur indifferently regarding all such aspects at once: multiple, they have no quantity; univocal, they apply to the real, the possible, and impossible; pre-individual, they are neither particular nor universal; impersonal and aconceptual, they are not material, mental, or useful; problematic, neither affirmative nor negative, they belong to no dimension of the proposition. Despite this neutrality, the Event also generates intelligibility in propositions and things, emitting singularities, producing series, and enveloping them.[35]

UNFOLDING TIME

How does this neutrality harmonize with the transcendental field's power to generate the sense of states of affairs and propositions? Both neutrality and generativity belong to a paradoxical element.

Because sense and nonsense coexist in this element, it remains neutral regarding all the oppositions of sense. Nonsense within the paradoxical element produces the problematic character of the transcendental field, for problems raise questions but remain neutral regarding answers. Ontologically, this problematic corresponds to an objective zone of indetermination, a singular aleatory point that envelopes other singularities. This singularity distributes points via the displacement and circulation of the paradoxical element between corresponding topological planes, generating an independent organization of sense. Occurring at the boundaries between topological planes, the paradoxical element circulates among things, between things and propositions, and within the proposition; it brings events together with language, attributing sense to things and expressing it in propositions. The paradoxical element acts as quasi-cause immanent to its effects: bifurcating, producing series, and organizing other points.

Multiple and diverse, the paradoxical element instantiates an abstract machine that diagrams the neutrality and generative power of the paradox.[36] Paradoxes install the problematic inherent in language, and generate the dimensions of the proposition. Rather than indicating epistemological failures, paradoxes mark positive ontological and logical effects. Paradoxes generate infinite regresses and vicious circles, i.e. series and envelopments. Although various determinations may fulfill the role of the paradoxical element which may circulate among diverse corresponding topological planes, semantic and set-theoretic paradoxes best exemplify the paradox's abstract diagram.

Propositions express sense, i.e. those conditions under which they can denote things or states of affairs. While propositions contain their sense a priori, a paradox arises if a proposition denotes its own sense. For then that proposition could have its sense only were it true. Yet, since the proposition's sense provides the conditions under which it could be true or false, its sense must be neutral regarding its truth or falsity; therefore sense belongs to propositions prior to their truth values. The paradox engenders a vicious circle. Such circles extend to two propositions, each denoting the sense of the other, and to more complex variations. Although no proposition can denote its own sense, another proposition can denote the sense of the first as long as the first does not denote the sense of the second. If this procedure does not relapse into a vicious circle it leads to an infinite regress, in serial form, where for every proposition p_j that denotes the sense s_i of a proposition p_i there will be another proposition p_k that denotes the sense s_j of p_j. This forms a homogeneous series of propositions (distinguished by degree or type) that necessarily subsumes two alternating heterogeneous series of propositions and senses. The paradoxical element instantiates this duality of intertwined heterogeneous series and circulates between such series linking corresponding topological planes.

The third synthesis is itself double. The paradoxical element effects both regressive and disjunctive syntheses. Incomplete by nature, this element splits apart, never appearing itself, but generating appearances on both its sides. A proposition that denotes its own sense is nonsense. This nonsense generates an infinite regress in which sense, endlessly pursued by nonsense at each level, continually reappears in a proposition of a higher type. At each stage the regress may be halted, splitting off a virtual double expressing the missing sense. This virtual double exceeds the proposition's dimensions, remaining neutral regarding oppositions within the dimensions of the proposition it doubles. This continuous excess marks the displacement of a mobile impossible object, the paradoxical element. Nonsense and sense are not alternative values like true and false, and do not exclude one another. The nonsense of the paradoxical element is the condition of, and so always co-present with, sense. Nonsense produces an excess of sense that remains hidden or appears in a proposition on a higher level. The paradoxical element advances in perpetual disequilibrium with respect to itself, a pure differenciation endlessly displacing itself.

When two propositions denote each other's sense, the sense of each determines an alternative into which the proposition with that sense enters, producing a vicious circle. In this way, the paradoxical element may belong to incompatible alternatives within the disjunctions it generates; the paradoxical element both generates disjunctions and envelops them. The paradoxical element coordinates heterogeneous series, circulating between them by displacing itself and so making them converge by making them endlessly diverge in multiple disjunctions. These coordinated series will include contradictory determinations that have sense even though only impossible objects could fulfill their denotations. This distinguishes their virtual being from the possible, which is governed by the principle of non-contradiction. Since they are neutral conditions continuously splitting into multiple disjunctions, paradoxes engender possibilities and the laws of non-contradiction and excluded middle, but are not subject to them.[37]

The syntheses of time instantiate the diagram of the paradox.

The third synthesis acts as an effective, because paradoxical, ontological principle expressing sense. Strictly nonsense, this synthesis
univocally expresses all sense at once in the empty form of Aion.
Affirming both sense and nonsense at once, the paradox goes in
both directions. The singular aleatory point traversing the line of
Aion forms the unequal boundary between topological planes, splitting virtual from actual, language from things, while articulating
sense on both sides. The continuous division of past and future
occurs as an event, univocally expressed by an infinitive verb.[38]
Since all sense emanates from its nonsense, the paradoxical element
effects a disjunctive synthesis. Instead of an analysis that refers
differences to a more encompassing identity, and comprehending
them through negation or opposition in exclusive disjunctions, it
gathers diverse events in an inclusive disjunction, making them
resonate across all disjuncts, positive singularities distanced from
one another by the paradoxical element's displacement.

   In continuous disequilibrium with itself, and so continuously
displaced, the third synthesis ramifies endlessly in the eternal return
of the paradoxical element with its unequal sides. Something new
arises because the continuous displacement of this nonsensical and
aleatory point traverses the line of Aion. This displacement occurs
between presents, continuously dividing into past and future,
forming the Event distributing and extending unities and identities
while simultaneously destabilizing them, an effect internal to its
cause.[39] Sets open on to wholes, and things lose their identities
through the coexistence of incompatible differences, a multiplicity
of senses communicating in the unfolding of the Event. The eternal
return selects this pure Event, the open whole in which all other
events communicate, affirming the differences between them by the
continuous distancing of past and future. This divergence and
distancing does not form a coherent world, but its transcendental
precursor.

   In the third synthesis of time this transcendental precursor
constitutes the objective problematic of passage. Perpetually
exchanging virtual and actual, it forms the crystal in which the
future expresses the perpetual exchange of virtual past and actual
present. Events expand at their edges, crystallizing past and present.
The third synthesis of disjunction generates this expansion by
producing, on one side, an infinite regress emitting the series of
presents functioning as the first, connective, synthesis of time and,
on the other side, a vicious circle enveloping them in the open whole
of the past, functioning as the second, conjunctive synthesis of

time. Besides producing and differenciating these dissymmetrical syntheses out of its paradoxical form, the third synthesis makes them resonate together and so generate the active synthesis – intelligible experience. In the first synthesis events unfold, past and future, between presents continuously displacing themselves, generating the continuous series of former presents. Without a place on the line of Aion, the paradox continuously displaces the instant along that line. However, since the paradox simultaneously generates two heterogeneous series resonating within every homogeneous series, each actual present coexists with its virtual past. This virtual past does not resemble or duplicate its present, but doubles it with the whole of its past. The virtual past conjoins and envelopes its former presents in a whole containing the totality of the past opening on to ever new presents.

Multiple envelopments of the series of presents in the virtual past not only arise from the heterogeneous split in the series of presents, but also condition the active synthesis of consciousness and representation. The paradox in the third synthesis continually generates the new, continually splitting past from present, distancing past and future, and, keeping the past open, generating the first and second syntheses. The paradox generates an infinite regress, instantiated in the series of presents and connected in the living present by the displacement of the paradoxical element. The paradox also generates a vicious circle, an enclosure of self-reference, providing the reflective dimension, the condition for the representation of temporal passage, by enveloping the series of presents in the whole of the past. What appears in experience arises from the continual circular exchange of virtual past and actual present in the crystal of time that makes them coexist in the present. This coexistence gives each present, once generated by the first synthesis, its sense, making it an object of denotation.[40] Moreover, this circular synthesis makes the series of presents coincide with their envelopment in the coexistent whole of the past, embedding presents in themselves by referring to them from the dimension of the virtual past. Once generated by the second synthesis, this envelopment constitutes the signification of former presents as past, and constitutes their possibility. The second synthesis distributes sense along homogeneous series of presents by doubling it with a heterogeneous series, each member of which contains the whole of the past. Further envelopment makes these heterogeneous series communicate in inclusive disjunctions manifesting an infinite series of reflections.[41]

In this way, the three passive syntheses generate the active synthesis, the intelligible organization of experience and representations. Within this active synthesis, the living present is given to significations and manifested subjects that distinguish its division into former presents and their possibilities.[42] This intelligible organization constitutes the phenomena described by phenomenology. However, rather than Husserl's individuating form or Heidegger's primordial unity, the multiple differenciation of the passive syntheses of time constitutes the dimensions of sense in propositions and things. This explains their genesis without raising any of the contents of this empirical field to transcendental status. Moreover, phenomenology itself gave an account of the constitution in experience of McTaggart's A- and B-series. The running-off of former presents as re-presented in recollection constituted McTaggart's A-series.[43] The passive syntheses condition this representation and recollection. B-theory, phenomenology, and Bergson agree that the quantification over actual times instantiates the B-series. However, phenomenology sees the B-series only as an aspect of objective time as it is given to consciousness, and Bergson argues that its quantitative multiplicity is an abstraction from concrete experience.[44] Within the active synthesis with its already constituted intelligible organization, actual time, the series of presents, forms a B-series. But the passive syntheses of time generate the active synthesis with its intelligible organization, including the possibilities of all quantity and measure. The active synthesis and the series it encompasses can never be completely actualized. As long as something new arises, the series of presents can never form a closed set, nor the living present of Chronos attain its own measure. The third synthesis is the neutral and generative precursor of the series of presents and of its intelligibility. This synthesis of the future, the eternal return of difference, unfolds the split between coexistent past and present, and generates something truly new.

Notes

I GREEK THOUGHT BEFORE ARISTOTLE

1 A standard source for more on the Presocratic philosophers and their mythographic predecessors is Kirk *et al.* 1974. A detailed examination of the arguments can be found in Barnes 1982. For a survey of Greek thought on time, see Lloyd 1976.
2 Hesiod 1982: 154–210.
3 Homer 1967: bk. 8, ll. 478–81.
4 Plato, *Apology*, 18b, 23d, in Plato 1961: 5, 9.
5 Kirk *et al.* 1974: 117ff.; translation modified.
6 Diels fragment 52: αιων παις εστι παιζων, πεσσευων παιδοs η βασιληιη.
7 Kirk *et al.* 1974: 273; translation modified.
8 In the dialogue Plato has Parmenides deduce the consequences of the assumptions both that there is a one and that there is not a one. Here, "there is a one" may be understood as referring to one thing that can be spoken or thought about, as opposed to many things (Plato 1961: 934ff.).
9 See Chapter IV, the section on "The nature of eternity".

II ARISTOTLE

1 Aristotle treats of time primarily in *Physics*, 4.10–4.14. On the question of the existence of time, see Chapter III, the section on "Skepticism and the existence of time."
2 *Physics*, 4.11, 218b 21–219a 1.
3 See the discussion of the soul in the next section.
4 Cf. Sorabji 1983: 74–78 for a detailed discussion of this issue and a defense of a verificationist interpretation.
5 Plotinus gives this criticism. See Chapter IV, the section on "Plotinus and Aristotle".
6 *Physics*, 2.1, 192b 1–32.
7 Cf. Sorabji 1983: 86–89.
8 *Physics*, 6.1, 231b 6–9.
9 *Physics*, 4.11, 219b 22–220a6.
10 *Physics*, 4.11, 219b 12–16, 26–32.

11 *Metaphysics*, 3.5, 1002b 5–8.

12 Cf. Sorabji 1983: 46–51. For interpretations that treat Aristotle's now as in transition, see Owen 1976; Heidegger 1982; and Chapter XII, the sections on "Existence and temporality" and "Ecstasis and temporal unity." We shall return to the static/flowing distinction below.

13 *Physics*, 2.1, 193b 4, 2.2, 194a 14–15.

14 *Physics*, 4.14, 223b 11–12.

15 *Physics*, 4.13, 222b 6–7. At *Physics*, 8.1, 252a 14–19, Aristotle argues that motion could not have a beginning because it would not be natural for it to begin at one time rather than another. This argument is repeated throughout the history of thought about time. We shall see it elaborated by Leibniz and directed at a different target.

16 Callahan 1968: 86–87.

17 *On Interpretation*, 9, 18a–19b.

18 For a discussion of the way the ancients understood the issues, see Rist 1969: ch. 7. For contemporary discussions, see the essays on Aristotle's logic in Moravcsik 1968.

19 This point will be important throughout Part Two.

III GREEK THOUGHT AFTER ARISTOTLE: SKEPTICS, EPICUREANS, AND STOICS

1 Aristotle, *Physics*, 4.10, 217b 33–218a 3.

2 Sextus Empiricus 1990: 235–239.

3 This argument is anticipated by Aristotle. Later we shall see St. Augustine using it for different purposes.

4 Aristotle, *op. cit.*, 4.10, 217b 29–218a 30.

5 This argument has affinities with Zeno's arrow. (See Sorabji 1983: 332–334, 365–366; A. W. Moore 1990: 25–26.)

6 Epicurus 1926: 32–33.

7 Lucretius 1975, bk. 2: 207–251.

8 *Ibid.*: 103, 244.

9 Epicurus, *op. cit.*: 35.

10 Lucretius, *op. cit.*: 259–260; for a detailed discussion, see Deleuze 1990: 269–270.

11 *Ibid.*: 32–37, 440–482.

12 Maimonides 1963: I, 73. See also Duhem 1985: 338–340.

13 Duhem 1985: 338–351.

14 Sorabji, *op. cit.*: 384–402.

15 For more details, see Whitrow 1980: 204.

16 Stobaeus, *Eclogae*, I, 106.

17 Sambursky 1987c: 93ff., 104.

18 Stobaeus, *op. cit.*.

19 This suggestion is made in Rist 1969: 276–277.

20 Hawking and Ellis 1973: 3, 8, 358–359.

21 Rist, *op. cit.*: 278.

22 Stobaeus, *op. cit.*

23 Goldschmidt 1953: 39–44.

24 Stobaeus, *op. cit.*: I, 8, 42.

Notes

IV NEOPLATONISM AND THE END OF THE ANCIENT WORLD

1 Plotinus 1991: 213–232.
2 See Callahan 1968: 94–102 for a discussion of the differences between Aristotle and Plotinus in criticizing their predecessors.
3 See Chapters 5 and 6.
4 Here is the germ of the notion that it is a mistake to treat time quantitatively like space. This notion will be developed only much later, by Hermann Lotze, Samuel Alexander, and especially Henri Bergson. See Chapter X, the section on "Translation and reduction"; and Chapter 13.
5 Sorabji 1983: 89.
6 These non-temporal senses of "before" and "after" are not attributes of motion as are the pre-temporal senses found in Aristotle.
7 See Chapter I, the section on "The moving image of eternity: Plato."
8 The main sources on Late Neoplatonism's views of time are given in Sambursky and Pines (1971) and Simplicius (1992). See also comments in Sorabji (1983) *passim*. This section is heavily indebted to Sambursky and Pines' introduction.
9 See Chapter II, the section on "The now."
10 The distinction between the two ways of describing time will become important again in the twentieth century in the work of John Ellis McTaggart. (See Chapter IX.)
11 Sorabji *op. cit.*: 43n.
12 "Just as the parts of separate things do not overlap because of space, the occurrence of the Trojan war does not become mixed up with that of the Peloponnesian war because of time, nor in a man's life does the state of a new-born become mixed up with that of a youth" (Damascius; quoted by Simplicius *Physics*, 775, 16; in Sambursky 1987b: 14).
13 See Sorabji *op. cit.*: 52–63 for a detailed discussion of this innovation.

V ANTICIPATIONS OF MODERNITY

1 Augustine 1988: bk. XII, chs. 13–14.
2 Augustine 1961: bk. VII, ch. 20.
3 For a detailed and fascinating account of the reasons for the impact of Christianity on the Roman Empire of the fourth century, see Cochrane 1954.
4 See Chapter IV, the section on "The nature of eternity."
5 Augustine's account of time, as presented here, is primarily given in Augustine 1961: bk. IX.
6 See Chapter III, the section on "Skepticism and the existence of time."
7 This may well be the first use of a sound as a paradigm example of something with duration. The example becomes significant in the twentieth century because perceived sounds are primarily temporal rather than spatial. (See Chapter XI.)
8 This is Augustine's lament in Augustine 1961: bk. XI, ch. 29.
9 Augustine 1988b: bk. IV, ch. 18; bk. XV, ch. 26.

10 Augustine 1961: bk. XII, ch. 11; for a more detailed exposition, see R. Jordan 1955.

11 Augustine 1961: bk. XII, ch. 9

12 See Chapter VI, the section on "Absolute time: Newton".

13 Massingnon 1951; see also the discussion of Bonet and Canonicus in the section on "Transition to modernity," below.

14 Sharif 1966: 1113; see also Wolfson 1919.

15 See Duhem 1985 for a detailed account of these developments and selections from the original texts.

16 From *Quaestiones Quodlibetales* XI, as noted in Duhem 1985.

17 On virtual existence, see Duns Scotus 1975: Lectura I, d.2, n.272, and Glossary, p. 538; also Hyman and Walsh 1983: 622–624. Peter Aureol (1280–1322) elaborated the actualization of time by measurement (Duhem *op. cit.*).

18 Ockham's opponents in this matter include Franciscus de Marchia (1285–1328) and John Buridan (*c.*1300–1358).

19 Duhem 1985: 321–323.

20 Clagett 1959: 199–222.

21 For a more detailed study, see Clagett (1968) and the summary of these developments in Whitrow (1980: 120–128).

22 The Newtonian conception of absolute time is discussed in Chapter VI.

23 On the Benedictine schedule, see Zerubavel 1981: 31–69; see also Mumford 1934: 12–18.

24 For the history of the mechanical clock, see Cipolla 1978; Landes 1983.

25 Landes 1983: 72–76.

26 See Le Goff 1988: 39–41.

27 A similar debate continues in the twentieth century. (See Chapters X and XIII.)

VI ABSOLUTE AND IDEAL TIME

1 Descartes 1972: vol. II, 219.

2 Descartes 1972: vol. I, 242.

3 Galilei 1914: 155.

4 Barrow is extensively quoted in Burtt 1954: 150–161.

5 Newton 1962: 131–132. On the ontological status of time in Newton, see Carriero 1990: 124–128. Newton states his theological views in the General Scholium added to the second edition of his *Mathematical Principles of Nature* in 1713.

6 For the Neoplatonic origins of Newton's conception of time, see Koyré 1957: chs. VI, VII.

7 Alexander 1956: 40; for Newton's reply, see Newton 1952: 370–371.

8 Newton 1952: 8.

9 See Chapter V, the section on "Emerging perspectives."

10 Newton 1969: 17.

11 Capek (1988) takes irreversibility to be the core of the conception that time flows.

12 Boyer 1949: 187–223.

13 This claim presumes the correctness of Newton's laws of motion, which Newton takes to be empirically supported. For a discussion of this point, see Dieks 1988b.

14 Toulmin 1959.

15 Locke 1959: 238–256.

16 Odegard 1978.

17 Compare Aristotle's remarks on "those who sleep among the heroes in Sardinia" (see Chapter II, the section on "Time and change").

18 See Chapter III, the section on "The Epicureans: time atoms."

19 Berkeley 1977: 113–114. On Berkeley's theory of time and objections to it, see Macintosh 1978.

20 Leibniz's criticisms of absolute time can be found in his correspondences with Samuel Clark (Alexander 1956). For his critique of Locke, see Leibniz 1981.

21 On the principle of sufficient reason, see Leibniz 1902: 141, 258; Mates 1986: 154–162; Russell 1989: 30–39; Deleuze 1993: 41–58; on the identity of indiscernibles, see Leibniz 1902: 14–15; Mates 1986: 132–136; Russell 1989: 54–63; Deleuze 1993: 41–58.

22 Aristotle used a version of this argument to show that the universe could not have a beginning in time (see Chapter II, the section on "The scope of time").

23 This argument is given in Shoemaker 1969.

24 Leibniz 1956: 68.

25 On the distinction between real relational properties of things and the ideal relations abstracted from them, see Ishiguro 1972.

26 "there must be in the simple substance a plurality of conditions and relations, even though it has no parts" (Monadology 13; Leibniz 1902: 253).

27 On time and duration, see Spinoza 1905: 129–130; *The Ethics*, Pt II, Definition 5, and its explication in Spinoza 1992: 63.

28 *The Ethics*, Pt I, Proposition 15, Scholium, in Spinoza 1992: 42. This distinction warrants comparison with the views of the Stoics. (See Chapter III, the section on "The Stoics: continuous time".)

29 *The Ethics*, Pt I, Definition 6, in Spinoza 1992: 31. For the comparison with Neoplatonic emanation, see Wolfson 1958: vol. I, 331–369; and Hardin 1978.

30 Russell raises this problem in Russell 1989: 50–53.

31 Leibniz 1989: 178–181; also Deleuze 1993: 100–120.

VII KANT

1 Kant 1965: A39–41/B56–58.

2 *Ibid.*: A369ff., Bxvi–xviii.

3 *Ibid.*: A34/B50.

4 *Ibid.*: A19.

5 *Ibid.*: A31/B46.

6 *Ibid.*: A33/B49.

7 *Ibid.*: B69ff.

8 *Ibid.*: B71, B145.

9 *Ibid.*: A34/B50.

10 *Ibid.*: B153–156.
11 *Ibid.*: A98–99, A210/B255–256.
12 *Ibid.*: A32/B48–49, A152/B191.
13 *Ibid.*: A32/B47.
14 *Ibid.*: A32/B47–48; see also *ibid.*: A25/B40, A38/B406; Allison 1983: 43.
15 *Ibid.*: A32/B47.
16 *Ibid.*: A50/B74.
17 *Ibid.*: B161.
18 *Ibid.*: A89–90/B121–122.
19 Schrader 1951.
20 Kant 1965: B160.
21 *Ibid.*: B151.
22 This accepts Aristotle's point (in Chapter II, the section on "The now") that time cannot be made up of nows.
23 In this way Kant offers a transcendental version of Augustine's notion that all times must be given in the present (see Chapter IV, the section on "The nature of eternity"). On the unity of temporal moments, see Allison 1983: 160–164; and regarding the necessary perspective, see Friedman 1954.
24 See Kant 1965: A137/B176 ff. for Kant's account of the schematism of the categories.
25 *Ibid.*: A181/B224.
26 *Ibid.*: A187–8/B230/1.
27 See Chapter VI, the section on "Absolute time: Newton."
28 *Ibid.*
29 See Chapter II, the section on "The definition of time."
30 On the distinction between the actually or metaphysically infinite and the mathematical infinite or unlimited expansion, see A. W. Moore 1990.
31 Kant 1956: 126–128; see also Kant 1960: 32, 60–61.
32 Kant 1965: A248–260/B305–315.

VIII BEING AND BECOMING

1 See Chapter VII, the section on "The nature of time."
2 See Chapter VII, the section on "Temporal limitations." For Hegel's rejection of noumenal things in themselves, see Hegel 1989: 46–47; Hegel 1977: 97–98.
3 Kant 1965: A326–7/B383.
4 C. Taylor 1975: 226–227.
5 Hegel 1989: 234.
6 This recapitulates Plotinus' formula (see Chapter IV, the section on "The nature on eternity").
7 Hegel 1977: 27.
8 Hegel 1989: 105.
9 *Ibid.*: 85.
10 Hegel 1970: 33–40.
11 Kojève 1969: 159–160.
12 See Chapter II, the section on "The now."
13 Hegel 1970: 39.

14 Hegel 1977: 100.
15 Nietzsche 1968b: § 707.
16 *Ibid.*: §§ 331, 765.
17 *Ibid.*: § 708.
18 Deleuze 1983: 23–24.
19 See Chapter I, the section on "Time and justice: Anaximander."
20 Nietzsche 1986: 326, § 67.
21 See the next section.
22 Nietzsche 1968b: §§ 480–492, 552.
23 *Ibid.*: §§ 477, 480, 552, and 676.
24 Stambaugh 1987: 129.
25 Nietzsche 1974: § 84.
26 Nietzsche 1968b: § 617; Nietzsche 1968a: 430. This is similar to the Stoics' Aion (see Chapter 3, the section on "The Stoics: continuous time").
27 Nietzsche 1968a: 485–486; 1974: § 54; *Works* 10: 214.
28 Klossowski 1977: 111–112; Stambaugh 1987: 189–190.
29 Deleuze *op. cit.*: 43.
30 Nietzsche, *Works* 14: 418; quoted in Stambaugh 1987: 132.
31 See Chapter IV, the section on "The nature of eternity."
32 Nietzsche 1968b: 339, §§ 635, 660.
33 Nietzsche 1968b: 471, n. 7. On the qualities of forces, see *ibid.*: § 564.
34 *Ibid.*: § 551; see also Deleuze, *op. cit.*: 64–68.
35 Nietzsche 1968b: § 1067.
36 *ibid.*: § 55.
37 Danto (1965) explores this point, but thinks it refutes Nietzsche's argument, since he interprets that argument as an attempt to legitimate eternal return with natural scientific methods.
38 Nietzsche 1968b: § 1062.
39 Stambaugh 1972: 103. For Newton's reliance on God, see Chapter VI, the section on "Absolute time: Newton."
40 Nietzsche 1968a: 269–70. Plato attributes to Parmenides the notion that time does not pass bit by bit but, rather, is the dividing moment that becomes both older and younger than itself (Plato 1961: 934–935; and Chapter I, the sections on "Questioning the reality of time: Parmenides" and "The moving image of eternity: Plato."
41 Nietzsche 1966: § 56.
42 Nietzsche 1968a, *op. cit.*
43 Nietzsche 1968b: § 617.
44 See Chapter 1, the section on "The moving image of eternity: Plato"; Nietzsche 1968a; 1968b: 206: "Every ring strives and turns to reach itself again."
45 Nietzsche 1966: § 150.
46 Stambaugh 1972: 123–126.
47 Nietzsche, *Works* 18: 320; quoted in Stambaugh 1987: 36, 202.
48 Deleuze, *op. cit.*: 50.
49 Nietzsche, Nachlaß, XII: 65; translation in Stambaugh 1972: 55.
50 Klossowski 1977: 111.
51 Stambaugh 1987: 65; Klossowski 1977: 112.
52 Nietzsche 1974: § 54; 1968b: § 853; 1968a: 485–486.

53 Deleuze, *op. cit.*: 26.
54 Nietzsche 1968b: §§ 639, 1067.
55 Nietzsche 1986: 325 § 61; 1968a: 500–1.
56 Nietzsche 1968b: § 1032.
57 *Ibid.*: § 1032.
58 Deleuze, *op. cit.*: 175–189.
59 *Ibid.*: 71.

IX MCTAGGART'S PROBLEM

1 McTaggart 1908.
2 McTaggart 1927: vol. II, 9–31.
3 Static and dynamic time are discussed in Chapter IV, the section on "Late Neoplatonism."
4 Russell 1903, § 442: 469–471.
5 For a historical synopsis of the idea of a specious present, see Mabbot 1967.
6 McTaggart 1927: vol. II, 19. In the 1908 essay McTaggart expressed no such preference for relations over qualities.
7 Phenomenology endeavors to face up to this difficulty (see Chapter XII, the section on "Constituting the absolute temporal flow").
8 For example, see Hall 1934: 339; Mellor 1981: 92.
9 See Chapter IV, the section on "Late Neoplatonism," especially the diagram of the Pseudo-Archytas (Figure 4).
10 See Chapter X, the section on "Translation and reduction."
11 Broad 1959; see also Chapter I, the section on "The moving image of eternity: Plato."
12 Schlesinger 1980: 33.
13 *Ibid.*: 136.
14 E.g. *ibid.*: 30.
15 Broad, *op. cit.*; Williams 1967; Pears 1956; Smart 1963: ch. vii.
16 This answer is given in Broad 1938. A-theories differ regarding whether to become present is to come into existence (see Mundle 1959; see also Chapter X, the section on "Time and existence").
17 Prior 1958.
18 Schlesinger, *op. cit.*: 31–32; Oaklander and Smith 1994: essays 20, 21, 22, 25. The dispute between Oaklander and Schlesinger, for example, revolves around the complex relations between the two-time-series account of the now's rate of change and Schlesinger's connecting of the now with actuality. On time and existence, see Chapter X, the section on "Time and existence."
19 E.g. Williams, *op. cit.*
20 Mellor 1981: 114–116.
21 Marhenke 1935. On the now as event, see Chapter X, the section on "Time and existence."
22 *Ibid.*; Mundle 1959; see also Chapter X, the section on "Translation and reduction."
23 Williams, *op. cit.*: 113.
24 See Chapter VII, the section on "The nature of time," for a similar point used for different purposes.

25 Mellor 1981: 104–107, 134–139.
26 Chapter X, the section on "Translation and reduction."
27 *Ibid.*: 119.
28 Strong 1935; Smart 1956.
29 Findlay 1967; Rankin 1958; Mellor, *op. cit.*: 132–134.
30 On these matters, see Shorter 1984 and Stebbing 1936.
31 Gotshalk 1930; Broad 1938.
32 Broad 1938; see also Prior 1958; Christensen 1974.
33 Prior 1958, 1968: ch. I; see also Pears 1956; Christensen 1974.
34 This point is later taken up in Lowe 1987a.
35 Mellor, *op. cit.* On the eliminability of tense, see his *ibid.*: ch. 10.
36 Mellor, *op. cit.*: ch. 6; Shorter, *op. cit.*
37 Schlesinger, *op. cit.*
38 Schlesinger, *op. cit.*: 56–58.
39 Prior 1967: 5–6; Schlesinger, *op. cit*: 51–52; Christensen, *op. cit.*
40 Rankin 1981.
41 Dunne 1958, criticized in Broad 1938.
42 This strategy was first suggested in Stebbing 1936 and elaborated in Smith 1994a.
43 This baroque example, typical of these debates, comes from Oaklander 1994. Smith acknowledges that presentness predominates in this analysis (Smith, *op. cit.*: 205).

X TENSE AND EXISTENCE

1 A version of this argument is presented in Smith 1994b.
2 Quine 1976: 145–148; Smart 1963.
3 Mink 1960.
4 Wisdom 1929; Stebbing 1936.
5 Gale 1964 attributes the compatibility of A-series representations to a rule of use rather than to their asserted content.
6 Pears 1956.
7 Lowe 1992.
8 Gale 1966.
9 Gale 1962; 1968: 40–47. Gale restricts the criterion so that it does not apply to sentences that are necessarily true or necessarily false, or to sentences that are not freely repeatable because they contain a non-temporal indexical expression.
10 See Chapter IX, the section on "The status of the paradox"; Pears 1956; Christensen 1974; while Smart (1963) defends temporal predicates over tensed verbs.
11 Quine, *op. cit.*.
12 Davidson 1984. For more see the next section on "Translation and reduction."
13 See Chapter IX, the section on "McTaggart's argument for the unreality of time."
14 Gale 1966; 1967: 92–95.
15 Gale 1966.
16 Gale 1967: 96.
17 Stebbing, *op. cit.*.

18 *Ibid.*
19 *Ibid.*; Broad 1938; Mackay 1935; Wisdom 1929; Gale 1963; Schlesinger 1980.
20 Gale 1963.
21 See the section on "Experiencing time," below.
22 Quine, *op. cit.*.
23 For versions of the date analysis, see *ibid.*; Russell 1903: 471; Goodman 1951: 295 ff.
24 Prior 1959.
25 Quine, *op. cit.*; Smart 1963: ch. VII.; see also the Excursus section below.
26 Whether a particular representation is a token of a sentence or of a judgment matters little for our purposes.
27 Smart 1963: 138–139.
28 Reichenbach (1966: §§50–51), Ayer (1956: 57–58) and Smart (1963: 133–134) give versions of the token-reflexive analysis, but this analysis has become a staple pervading the literature.
29 Schlesinger, *op. cit.*: 33; Gale 1963: 352–353.
30 Gale 1968: 49.
31 Schlesinger, *op. cit.*: 28–30.
32 Gale 1964; Oaklander and Smith *op. cit.*
33 Mayo 1950.
34 Perry 1979; regarding the ineliminable indexical element in A-series language, see Lowe 1987a.
35 Mellor 1981: 73–88.
36 Prior, *op. cit.*
37 Gale 1962; Mellor, *op. cit.*
38 Findlay 1967; 143–162; Smart 1963: 139 ff.; Gale 1968: 31–33.
39 See, for example, Marcus 1961; Donnellan 1966; Kripke 1971.
40 For the New B-theory, see Mellor, *op. cit.*; Le Poidevin 1991; Le Poidevin and MacBeath 1993; Oaklander and Smith, *op. cit.*
41 Mellor, *op. cit.*: 98–102. The argument here is a version of McTaggart's argument, stated ontologically rather than linguistically, that the A-series is contradictory where variation in the truth values of tensed sentence tokens conflict with the constancy in the truth values of sentence types.
42 Mellor, *op. cit.*: 73–88.
43 Einstein 1952.
44 Named after H. A. Lorentz, who suggested them in Lorentz *et al.* 1952 and other papers.
45 For dissenting views, see Stein 1968; McCall 1976; Dieks 1988.
46 For causation and the theory of relativity, see Newton-Smith 1980: ch. VIII. For the view that the direction of causation determines the direction of time, see Stebbing, *op. cit.*; Schlesinger, *op. cit.*; Mellor, *op. cit.*
47 See Chapter VI, the section on "Absolute time: Newton."
48 Newton-Smith 1980: ch. VIII; Mellor, *op. cit.*: 66–72.
49 See Chapter II, the section on "The definition of time."
50 See Chapter III, the section on "The Epicureans: time-atoms."
51 Stebbing, *op. cit.*; Smart 1956.
52 Mundle 1959.

53 Broad, *op. cit.*; Paul Marhenke 1935.
54 Williams 1967.
55 Mellor, *op. cit.*: 30.
56 Prior 1967: ch. VIII.
57 See Chapter VI, the sections on "Absolute time: Newton" and "Locke and empiricism."
58 Mackay, *op. cit.*: 180.
59 See the section on "Experiencing time," below.
60 Broad, *op. cit.*; Mundle, *op. cit.*
61 Rankin 1958; Hall 1934.
62 Mellor, *op. cit.*: 31.
63 Schlesinger, *op. cit.*: 34–36. Rankin, *op. cit.*.
64 Smart 1963: 131–132.
65 See Chapter III, the section on "The Epicureans: time atoms"; Chapter VI, see the sections on "Absolute time: Newton" and "The activity of substance: Spinoza and Leibniz."
66 For discussions of how conscious understanding stands in relation to present events, see Dummett 1978 and McDowell 1978.
67 Mellor, *op. cit.*: ch. 3. This view claims that it can accept the ineliminability of indexical devices such as tense because it depends on what is referred to by such expressions rather than on how their reference is secured. Opponents reply that this makes the unwarranted assumption that time can be viewed *sub specie aeternitatis* (Lowe 1987a, 1987b; Le Poidevin and Mellor 1987).
68 Smart 1963: 142–143.

XI PHENOMENOLOGY OF TIME

1 For a history of these writings, see Husserl 1991: XI–XVIII.
2 *Ibid.*: 336.
3 Husserl 1982: 283.
4 *Ibid.*: 44–45, 39. Here, experience means given in intuition.
5 Husserl 1981.
6 See Chapter X, the section on "Experiencing time."
7 See Chapter XII, the section on "Constituting transcendent temporal objects."
8 On phenomenological constitution, see Sokolowski 1964.
9 Cairns 1968.
10 This objection applies to similar accounts supported by B-theory (see Chapter X, the section on "Experiencing time").
11 Husserl 1991: 11–15 summarizes this account. Note the affinity with Augustine's solution to the paradox of time's existence. (See Chapter V, the section on "Time and human personality: Augustine").
12 *Ibid.*: 16–20.
13 Intentionality is so central to phenomenology that discussions of it pervade the whole of the literature. (See Husserl 1982: 73–75; Heidegger 1985: 27–47; Cairns 1959/60; Sokolowski 1974; for the importance of intentionality, see R. W. Jordan 1974.)
14 In his later work Brentano corrected this confusion (see Kraus 1976).

Notes

15 These intendings provide a perspective that unifies temporal objects. Past, present, and future, then, do not form a series, as McTaggart claimed (see Chapter IX, the section on "McTaggart's argument for the unreality of time").

16 Husserl 1991: 25–27.

17 *Ibid.*: 30–31.

18 *Ibid.*: 29.

19 *Ibid.*: 33–36.

20 *Ibid.*: 40–42.

21 *Ibid.*: 37–38.

22 *Ibid.*

23 This is the point made by A-theory (see Chapter X, the section on "Time and existence").

24 *Ibid.*: 55–57. The fulfillment of protentions is always contingent, and a recollection that does not treat earlier anticipations of something later as problematic would misrepresent its object (see Chapter XII, the section on "Ecstasis and temporal unity").

25 For more on phenomenology and McTaggart's paradox, see Chapter XII, the section on "Constituting the absolute temporal flow."

26 *Ibid.*: 44–46.

27 *Ibid.*: 47–49.

28 *Ibid.*: 50–52.

29 *Ibid.*: 93.

30 Husserl 1982: 94. This does not include ideal objects, which are indeed transcendencies that can be adequately given.

31 Husserl 1991: 88–89.

32 *Ibid.*: 89–92.

33 *Ibid.*: 119–120.

34 *Ibid.*: 80–84.

35 This will be discussed further in Chapter XII, in the section on "Constituting the absolute temporal flow."

36 See Chapter XII, the section on "Constituting transcendent temporal objects."

37 Husserl 1991: 84–85.

38 In this way, time is a boundary condition on phenomena. Time constituting consciousness will be discussed in Chapter XII, in the section on "Constituting the absolute temporal flow."

39 Derrida 1973: 60–69. For analysis and critique of these arguments, see Wood 1989; Brough 1993.

40 Husserl 1991: 40–43.

41 *Ibid.*: 122.

42 *Ibid.*: 355–356.

43 See Chapter II, the section on "The now."

XII TRANSCENDENCE AND EXISTENCE

1 Husserl 1991: 73–75. On objectivation, see note 18 in the section on "Constituting the absolute temporal flow," below.

2 See Chapter 7, the section on "The function of time."

3 Husserl, *op. cit.*: 80.

4 *Ibid.*: 113–114.

5 *Ibid.*: 95–99.

6 Husserl 1991: 66–71.

7 Husserl 1982: 359 ff.

8 Husserl 1991: 90–92. Heidegger will reject this claim.

9 *Ibid.*: 73–75.

10 *Ibid.*: 101–103, 133–137.

11 See Chapter XI, the section on "Constituting immanent content."

12 See Chapter IX.

13 McTaggart 1908: 474.

14 Husserl 1991: 98.

15 *Ibid.*: 80. On passive syntheses, see Husserl 1970: 77–80.

16 Husserl 1991: 78.

17 *Ibid.*: 118.

18 To objectivate something gives it the categorial form "this-here," so that other aspects of its being (sense) can attach to it (Husserl 1982: 27–28).

19 Hegel also sought the unity of time in absolute subjectivity, but thought the absolute subject was universal, having the form of the concept (see Chapter VIII, the section on "Dialectical movement: Hegel"). For Husserl, however, absolute consciousness does not have the form of the concept.

20 Husserl 1991.

21 *Ibid.*: 79.

22 Mackay also makes this point (Mackay 1935: 180).

23 Husserl 1991: 74.

24 *Ibid.*: 99–101, 130–133.

25 See Chapter XI, the section on "Constituting immanent content."

26 Husserl 1991: 123.

27 Husserl 1982: 17–23.

28 Heidegger 1982: 227–228.

29 "Handy" translates the German "*zuhanden*" (*Ibid.*: 162–170).

30 "At-hand" translates the German "*vorhanden*" (*Ibid.*: 108–109).

31 *Ibid.*: 161–173.

32 *Ibid.*: 257; on Aristotle, see Chapter 2, the section on "The definition of time."

33 Heidegger 1982: 242–245.

34 *Ibid.*: 247.

35 See Chapter XIII, the section on "Pure multiplicities."

36 Heidegger 1982: 249.

37 See Chapter II, the section on "The scope of time."

38 Heidegger 1982: 252.

39 See Chapter II, the section on "The definition of time."

40 Analytic philosophers recognized a similar practical orientation (Chapter 10, see the section on "Translation and reduction").

41 Heidegger 1982: 261–264.

42 *Ibid.*: 259–260.

43 The Greek εκστατικον means to place beyond, to step outside itself.

44 Here "temporality" translates the German "*Zeitlichkeit*," as opposed to "*Temporalität*," which Heidegger uses for the understanding of *Zeitlichkeit* (Heidegger 1982: 228).

45 *Ibid.*: 266–268.
46 *Ibid.*: 268–271.
47 *Ibid.*: 275–276.
48 Heidegger also mentions "for-which" and "for-the-sake-of" relations (*Ibid.*: 293, 295).
49 *Ibid.*: 64, 170, 276; Heidegger 1962: 32–33.
50 Heidegger 1982: 296.
51 *Ibid.*: 272; see also Chapter III, the section on "Skepticism and the existence of time"; Chapter IX; and Chapter XII, the section on "Constituting the absolute temporal flow."
52 Heidegger 1982: 319–320.
53 *Ibid.*: 305–312.
54 *Ibid.*: 325.
55 *Ibid.*: 307.
56 In German, *"Ereignis"* and *"es gibt,"* respectively (Heidegger 1972: 17–19).
57 Wood 1989: 258.

XIII MULTIPLICITY AND VIRTUALITY

1 Ideal spaces are always abstract, and Bergson includes actual space as one instance of ideal space.
2 Bergson 1960: 75–85, 120–123. Bergson writes about numbers in terms of unifying mental acts, but that is not essential to his point (Lacey 1989: 18).
3 Bergson 1960: 98–99.
4 *Ibid.*
5 Bergson 1968: 207–217.
6 F. C. T. Moore 1996: 14–17.
7 "Either therefore the reality which underlies space must form a discrete manifold or we must seek the ground of its metric relations outside it, in the binding forces which act on it" (G. B. R. Riemann; quoted in Kline 1972: 893). The Riemannian metric for spatial distance is given by the constants g_{ij}, for a space of three dimensions, in equations of the form: $gds^2 = g_{11}dx^2 + g_{12}dxdy + g_{13}dxdz + g_{21}dydx + g_{22}dy^2 + g_{23}dydz + g_{31}dzdx + g_{32}dzdy + g_{33}dz^2$. In Euclidean space $g_{11} = g_{22} = g_{33} = 1$ and the other constants are zero (Boyer 1985: 589). See Chapter XIV regarding singularities-events which generate such metrics.
8 Bergson 1968: 19, 206; Deleuze 1988: 39.
9 Bergson 1988: 206–207; 1968: 21; Capek 1991: 56–70.
10 Contra Lacey, *op. cit.*: 25.
11 Bergson 1988: 71; 1960: 83–84.
12 Bergson 1968: 23.
13 *Ibid.*: 107–125.
14 *Ibid.*: 23–25, 113–123.
15 The actualization of the virtual bears a close kinship with the expression of Spinozan substance (see Chapter VI, the section on "The activity of substance: Spinoza and Leibniz").
16 *Ibid.*: 26–28, 107–110; Deleuze, *op. cit.*: 96–97.

17 Bergson 1955: 54–55; 1968: 80–81, 104. On Platonic metaphysics, see Chapter I, the section on "The moving image of eternity: Plato." Objectors often fail to see how radical is this reversal of flux and immobility. Such objections tend to presuppose the ancient priority of stability over flux, and then fault Bergson for violating that priority (e.g. Strong 1935: 62–80; for similar arguments, see Williams 1967: 101–105; R. Taylor 1955).
18 Bergson 1955: 24, 27.
19 Contra Lacey, *op. cit.*: 56.
20 Bergson 1960: 90–91, 107–109.
21 On quantification in B-theory, see Chapter X, the sections on "Time and existence" and "Experiencing time."
22 Bergson 1955: 29–31, 47–49; 1988: 149.
23 Bergson 1960: 110–115.
24 *Ibid.*: 106–110; Bergson 1965: 49–50; 1968: 10, 20.
25 Bergson 1960: 42–43; 1968: 14–16.
26 Bergson 1988: 56, 68–69, 228, 233.
27 *Ibid.*: 158–159; 1960: 128–139.
28 Bergson 1955: 37–45; 1968: 17–18. Heidegger also advocated the pragmatic constitution of the time line (see Chapter XII, the section on "Existence and temporality"). Pragmatists like E. A. Burtt and G. H. Mead also advanced a view attributing the representation of time to practical needs (see Hall 1934). These positions reverse the B-theory view that pragmatic concerns explain the use of A-language.
29 Bergson 1955: 36–37.
30 *Ibid.*; see also Bergson 1968: 31–32, 51–52; 1944: 179.
31 Bergson 1988: 142.
32 *Ibid.*: 183–185.
33 Bergson 1960: 127–128.
34 Bergson 1955: 34; 1968: 30.
35 Bergson 1988; 1944; 1965.
36 For a detailed account of Bergson's intuitive method, see Deleuze, *op. cit.*: ch. I.
37 Bergson 1955: 10–11, 112 ff.; 1968: 91–92, 103–105.
38 Bergson 1988: 133–142.
39 *Ibid.*: 142–143, 150–151.
40 *Ibid.*: 144–149.
41 Bergson 1955: 40; 1965: 48–49. This point was also central for Husserl (see Chapter XI).
42 Bergson 1988: 133–134, 148–149; 1968: 87; Deleuze, *op. cit.*: 54–56. For appeals to traces to explain passage and temporal direction, see Smart 1963 and Mellor 1981: 168–171.
43 Bergson 1988: 151–155.
44 *Ibid.*; Bergson 1975: 157–160. This, again, shows the impropriety of portraying *durée* as if it were McTaggart's A-series.
45 *Ibid.*
46 Bergson 1968: 69–70.
47 Bergson 1988: 103–105, 161–172.
48 In creative evolution, for example, in which different lines of actualization produce similar organs (Bergson 1944: 60–62).

49 On the ontology of *durée*, see Bergson 1944: 12–14; 1955: 45–46; 1965: 44–51. On emanation, see Chapter IV and Bergson 1968: 87, 110, 123–124.
50 Bergson 1988: 120, 129–131, 168–169. Bergson (1988: 92–99, 126–128, 174–177) supports these claims with experimental results regarding psychic blindness and aphasia.
51 Bergson 1965: 46–47.
52 On the special theory of relativity, see the "Excursus" in Chapter X. Bergson's major treatment of relativity is found in Bergson 1965.
53 A complete account of these issues must also consider the general theory of relativity, which treats accelerated motion. The standard physical account accepts the plurality of times, but rejects the inconsistency on the grounds that Paul must turn around, i.e. undergo an acceleration, to share a coordinate system in which to compare his age with Peter's. However, all this acceleration does is demonstrate the need for Paul to change coordinate systems to return to his starting point. For an assessment of Bergson's views on these matters, see Capek 1991: 296–323; Gunter 1969: 123–250.
54 Bergson 1965: 64–66, 70–86. Stebbing (1936) also makes this point.
55 For this equation, see the "Excursus" in Chapter X. On the difference in sign, see Mellor, *op. cit.*: 69.

XIV BECOMING-TIME

1 The distaff line is the genealogical line of descent that does not carry the family name, thus connecting seemingly isolated figures. In his discussions of time, Deleuze selects motifs from the Stoics, Duns Scotus, Spinoza, and Nietzsche, as well as Bergson, and incorporates insights from the Epicureans, Leibniz, and Kant, who do not belong to the main line of descent.
2 Husserl 1982: 287.
3 Deleuze 1991: 108–109; 1990: 98–99, 105–106; Deleuze and Parnet 1987: vii–viii.
4 Deleuze 1994: 19–20.
5 *Ibid.*: 70–79.
6 See Chapter XI, the section on "Recollection and time consciousness"; Deleuze 1994: 79–80.
7 There are, therefore, four syntheses of time, the three passive syntheses and the active synthesis of consciousness they produce.
8 Deleuze extracts these paradoxes from Bergson. (See Chapter XIII, the section on "Virtual coexistence").
9 Deleuze 1994: 81–82.
10 On parts and wholes, see Chapter XIII, the section on "Virtual coexistence."
11 *Ibid.*: 208; see also Deleuze 1988: 96.
12 Deleuze 1986a: 16–17; 1995: 55. This adapts Bergson's distinction between quantitative and qualitative difference (see Chapter XIII, the section on "Pure multiplicities").
13 See Chapter XIII, the section on "Virtual coexistence."
14 Deleuze 1989: 69–83.

15 See Chapter VIII, the section on "Nietzsche: eternal return."

16 See Chapter III, the section on "The Stoics: continuous time."

17 Deleuze 1989: 61–65, 164–168.

18 *Ibid.*: 58–61.

19 Plato recognized this aspect of events in acknowledging that the universe cannot be in time, except by becoming both older and younger than itself (see Chapter I, the section on "The moving image of eternity: Plato").

20 In this way, Aion is similar to the Epicurean clinamen (see Chapter III, the section on "The Epicureans: time-atoms").

21 Deleuze 1994: 89.

22 See Chapter III, the section on "The Stoics: continuous time;" Deleuze 1990: 5, 58–64, 77, 162–168.

23 Although, mathematically, singularities are points where functions are undefined, e.g. points of inflection and bifurcation, as events, singularities are embodied in critical points: points of fusion, boiling, condensation, precipitation, etc. Deleuze 1990: 52–53. Singularities-events condition all continuous measurements by grounding their metrics.

24 Deleuze 1990: 66–67; 1994: 119–124.

25 The unusual spelling indicates that differenciation does not merely classify identities, but is prior to the constitution of any identities.

26 See Chapter XIII, the section on "*Durée* and spatialization."

27 Deleuze (1994: 35–42) traces the lineage of univocal being from Duns Scotus, to Spinoza and Nietzsche.

28 Deleuze 1990: 94–99. The distinction between causes and quasi-causes evokes Spinoza's treatment of causes (see Chapter VI, the section on "The activity of substance: Spinoza and Leibniz"). Chains connecting causes with effects, which treat those effects as further causes, are empirically discovered. These are constituted within quasi-causal connections among singular points and presuppose their intelligible organization.

29 Therefore, events do not raise the problems of Chapter X, the sections on "Tenses and truth values" and "Translation and reduction."

30 Deleuze 1990: 37, 184–185.

31 *Ibid.*: 14–22. This is Lewis Carroll's paradox of what the tortoise said to Achilles.

32 *Ibid.*.

33 Kant 1965: 308–309.

34 Nietzsche called this will-to-power (see Chapter VIII, the section on "Nietzsche: eternal return").

35 Deleuze 1990: 100–105.

36 Deleuze and Guattari 1987: 141–148; Deleuze 1986b: 70–93.

37 Deleuze 1990: 28–41.

38 "To be" expresses being, or – better – "to become" expresses the Event in its continuous ramification, escaping Heidegger's nostalgia for unity (see Chapter XII, the section on "Ecstasis and temporality").

39 See Chapter VI, the section on "The activity of substance: Spinoza and Leibniz."

40 The present, then, has evidential status strictly within the intelligible arrangements of things. Its status is parallel to, but not dependent upon, representation, since the sense of the present and its representation are generated together. So, disregarding the conditions of its genesis, the active synthesis of consciousness divides between actual former presents and possible future presents, accepting phenomenology's retort to its critics (see Chapter XI, "Excursus").

41 For a detailed exposition of this genesis, see Deleuze 1990: 109–126.

42 See Chapter XI, the sections on "Recollection and time consciousness" and "Constituting immanent content"; Chapter XII, the section on "Constituting transcendent temporal objects."

43 Chapter XII, the section on "Constituting the absolute temporal flow."

44 Deleuze 1990; see also Chapter XIII, the section on "*Durée* and spatialization."

Bibliography

Alexander, H. G. (1956) *The Leibniz–Clarke Correspondence* (ed. and trans. H. G. Alexander), London: Philosophical Library, Inc.

Allison, Henry E. (1983) *Kant's Transcendental Idealism: An Interpretation and Defense*, New Haven, CT: Yale University Press.

Aristotle (1941) *The Basic Works of Aristotle* (ed. Richard McKeon), New York: Random House.

—— (1984) *The Complete Works* (ed. Jonathan Barnes), Princeton, NJ: Princeton University Press.

Augustine (1961) *Confessions* (trans. R. S. Pine-Coffin), London: Penguin Books.

—— (1988a) *City of God and Christian Doctrine*, in Philip Schaff (ed.) *Nicene and Post-Nicene Fathers of the Christian Church*, vol. II (trans. Marcus Dods), Grand Rapids, MI: Wm. B. Eerdmans Publishing Co.

—— (1988b) *On the Holy Trinity*, in Philip Schaff (ed.) *Nicene and Post-Nicene Fathers of the Christian Church*, vol. III (trans. Marcus Dods), Grand Rapids, MI: Wm. B. Eerdmans Publishing Co.

Ayer, A. J. (1956) *The Problem of Knowledge*, London: Macmillan.

Barnes, Jonathan (1982) *The Presocratic Philosophers*, London: Routledge.

Bergson, Henri (1944) *Creative Evolution* (trans. Arthur Mitchell), New York: Modern Library.

—— (1955) *An Introduction to Metaphysics* (trans. T. E. Hulme), Englewood Cliffs, NJ: Prentice-Hall, Inc.

—— (1960) *Time and Free Will: An Essay on the Immediate Data of Consciousness* (trans. F. L. Pogson), New York: Harper & Row.

—— (1965) *Duration and Simultaneity: With Reference to Einstein's Theory* (trans. Leon Jacobson), Indianapolis, IN: Bobbs-Merrill Company, Inc.

—— (1968) *The Creative Mind* (trans. Mabelle L. Andison), New York: Greenwood Press.

—— (1975) *Mind-energy: Lectures and Essays* (trans. H. Wildon Carr), Westport, CT: Greenwood Press.

—— (1988) *Matter and Memory* (trans. Nancy Margaret Paul and W. Scott Palmer), New York: Zone Books.

Bibliography

Berkeley, George (1977) *The Principles of Human Knowledge, With Other Writings*, Glasgow: William Collins Sons & Co. Ltd.

Boyer, Carl B. (1949) *The History of the Calculus and its Conceptual Development (The Concepts of the Calculus)*, New York: Dover Publications, Inc.

—— (1985) *A History of Mathematics*, Princeton, NJ: Princeton University Press.

Broad, C. D. (1938) *An Examination of McTaggart's Philosophy*, Cambridge: Cambridge University Press.

—— (1959) "A reply to my critics," in Paul A. Schilpp (ed.) *The Philosophy of C. D. Broad*, La Salle, IL: Open Court.

Brough, John B. (1993) "Husserl and the deconstruction of time," *Review of Metaphysics* 46: 503–536.

Burtt, Edwin Arthur (1954) *The Metaphysical Foundations of Modern Science*, Garden City, NY: Doubleday & Co. Inc.

Cairns, Dorion (1959/60) *Husserl's Theory of Intentionality*, vols. I, II, and III, transcriptions of tape-recorded lectures given at the Graduate Faculty of Political and Social Science, New School for Social Research, New York.

—— (1968) "An approach to phenomenology," in Martin Farber (ed.) *Philosophical Essays in Honor of Edmund Husserl*, Cambridge, MA: Harvard University Press.

Callahan, John F. (1968) *Four View of Time in Ancient Philosophy*, New York: Greenwood Press.

Capek, Milic (1961) *The Philosophical Impact of Contemporary Physics*, Princeton, NJ: D. Van Nostrand Company, Inc.

—— (1988) "What survives from the classical concept of absolute time," in P. B. Scheurer and G. Debrock (eds.) *Newton's Scientific and Philosophical Legacy*, Dordrecht: Kluwer Academic Publishers.

—— (1991) *The New Aspects of Time: Its Continuity and Novelties, Selected Papers in the Philosophy of Science*, Dordrecht: Kluwer Academic Publishers.

Carriero, John (1990) "Newton on space and time: comments on J. E. McGuire," in Phillip Bricker and R. I. Hughes (eds.) *Philosophical Perspectives on Newtonian Science*, Cambridge, MA: MIT Press.

Christensen, Ferrel (1974) "McTaggart's paradox and the nature of time," *Philosophical Quarterly* 24(97): 289–299.

Cipolla, Carlo M. (1978) *Clocks and Culture 1300–1700*, New York: W. W. Norton Co.

Clagett, Marshall (1959) *The Science of Mechanics in the Middle Ages*, Madison, WI: University of Wisconsin Press.

—— (1968) *Nicole Oresme and the Medieval Geometry of Qualities and Motions: A Treatise on the Uniformity and Difformity of Intensities Known as Tractatus de configurationibus qualitatum et motuum*, Madison, WI: University of Wisconsin Press.

Bibliography

Cochrane, Charles Norris (1954) *Christianity and Classical Culture: A Study of Thought and Action form Augustus to Augustine*, London: Oxford University Press.

Danto, Arthur C. (1965) *Nietzsche as Philosopher*, New York: Columbia University Press.

Davidson, Donald (1984) "Truth and meaning," *Inquires into Truth and Interpretation*, Oxford: Clarendon Press.

Deleuze, Gilles (1983) *Nietzsche and Philosophy* (trans. Hugh Tomlinson), New York: Columbia University Press.

—— (1986a) *Cinema 1: The Movement Image* (trans. Hugh Tomlinson and Barbara Habberjam), Minneapolis, MN: University of Minnesota Press.

—— (1986b) *Foucault* (trans. Seán Hand), Minneapolis, MN: University of Minnesota Press.

—— (1988) *Bergsonism* (trans. Hugh Tomlinson and Barbara Habberjam), New York: Zone Books.

—— (1989) *Cinema 2: The Time Image* (trans. Hugh Tomlinson and Robert Galeta), Minneapolis, MN: University of Minnesota Press.

—— (1990) *The Logic of Sense* (trans. Mark Lester with Charles Stivale), New York: Columbia University Press.

—— (1991) *Empiricism and Subjectivity: An Essay on Hume's Theory of Human Nature* (trans. Constantin V. Boundas), New York: Columbia University Press.

—— (1993) *The Fold: Leibniz and the Baroque* (trans. Tom Conley), Minneapolis, MN: University of Minnesota Press.

—— (1994) *Difference and Repetition* (trans. Paul Patton), New York: Columbia University Press.

—— (1995) *Negotiations* (trans. Martin Joughin), New York: Columbia University Press.

Deleuze, Gilles and Guattari, Felix (1987) *A Thousand Plateaus: Capitalism and Schizophrenia* (trans. Brian Massumi), Minneapolis, MN: University of Minnesota Press.

Deleuze, Gilles and Parnet, Claire (1987) *Dialogues* (trans. Hugh Tomlinson and Barbara Habberjam), New York: Columbia University Press.

Derrida, Jacques (1973) *Speech and Phenomena and Other Essays on Husserl's Theory of Signs* (trans. David B. Allison), Evanston, IL: North-western University Press.

Descartes, René (1972) *The Philosophical Works of Descartes*, vols. I and II (trans. Elizabeth Haldane and G. R. T. Ross), Cambridge: Cambridge University Press.

Dieks, D. (1988a) "Special relativity and the flow of time," *Philosophy of Science* 55: 456–460.

—— (1988b) "Newton's conception of time in modern physics and philosophy," in P. B. Scheurer and G. Debrock (eds.) *Newton's Scientific and Philosophical Legacy*, Dordrecht: Kluwer Academic Publishers.

Donnellan, Keith (1966) "Reference and definite descriptions," *Philosophical Review* 75: 281–304.

Duhem, Pierre (1985) *Medieval Cosmology: Theories of Infinity, Place, Time, Void, and the Plurality of Worlds* (ed. and trans. Roger Ariew), Chicago, IL: University of Chicago Press.

Dummett, Michael (1978) "The reality of the past," *Truth and Other Enigmas*, London: Duckworth.

Dunne, J. W. (1958) *An Experiment with Time*, London: Faber & Faber Ltd.

Duns Scotus, John (1975) *God and Creatures: The Quodlibetal Questions* (trans. Felix Alluntis and Allan B. Wolter), Princeton, NJ: Princeton University Press.

Einstein, Albert (1952) "On the electrodynamics of moving bodies," in H. A. Lorentz, A. Einstein, H. Minkowski and H. Weyl, *The Principle of Relativity: A Collection of Original Memoirs on the Special and General Theory of Relativity*, New York: Dover Publications, Inc.

Epicurus (1926) *Epicurus: The Extant Remains* (trans. and notes Cyril Bailey), Westport, CT: Hyperion Press.

Findlay, J. N. (1967) "Time: a treatment of some puzzles," in Richard M. Gale (ed.) *The Philosophy of Time: A Collection of Essays*, Garden City, NY: Anchor Doubleday & Company, Inc.

Frankfurt, Harry G. (1972) *Leibniz: A Collection of Critical Essays*, Notre Dame, IL: University of Notre Dame Press.

Friedman, Lawrence (1954) "Kant's theory of time," *Review of Metaphysics* VII, March: 379–388.

Gale, Richard M. (1962) "Tensed statements," *Philosophical Quarterly* 12: 53–59.

—— (1963) "A reply to Smart, Mayo and Thalberg on 'Tensed statements'," *Philosophical Quarterly* 13: 351–356.

—— (1964) "Is it now now?," *Mind* LXXIII: 97–105.

—— (1966) "McTaggart's analysis of time," *American Philosophical Quarterly* 3(2): 145–152.

—— (1967) (ed.) *The Philosophy of Time: A Collection of Essays*, Garden City, NY: Anchor Doubleday & Company, Inc.

—— (1968) *The Language of Time*, London: Routledge & Kegan Paul.

Galilei, Galileo (1914) *Dialogues Concerning Two New Sciences* (trans. Henry Crew and Alfonso De Salvio), New York: Dover Publications, Inc.

Goldschmidt, Victor (1953) *Le Système stoïcien et l'idée de temps*, Paris: Librairie Philosophique J. Vrin.

Goodman, Nelson (1951) *The Structure of Appearance*, Cambridge, MA: Harvard University Press.

Gotshalk, D. W. (1930) "McTaggart on time," *Mind* 39: 26–42.

Gould, Josiah B. (1970) *The Philosophy of Chrysippus*, Leiden: E. J. Brill.

Gunter, P. A. Y. (ed.) (1969) *Bergson and the Evolution of Physics*, Knoxville, TN: University of Tennessee Press.

Hall, Everett W. (1934) "Time and causality," *Philosophical Review* XLIII(4), July: 333–351.

Hardin, C. L. (1978) "Spinoza on immortality and time," in Robert W. Shahan and J. I. Biro (eds.) *Spinoza: New Perspectives*, Norman, OK: University of Oklahoma Press.

Hawking, S. W. and Ellis, G. F. R. (1973) *The Large Scale Structure of Space–Time*, Cambridge: Cambridge University Press.

Hegel, G. W. F. (1970) *Hegel's Philosophy of Nature* (trans. A. V. Miller), Oxford: Clarendon Press.

—— (1977) *Hegel's Phenomenology of Spirit* (trans. A. V. Miller), Oxford: Oxford University Press.

—— (1989) *Hegel's Science of Logic* (trans. A. V. Miller), Atlantic Highlands, NJ: Humanities Paperback Library.

Heidegger, Martin (1962) *Being and Time* (trans. John Macquarrie and Edward Robinson), New York: Harper & Row.

—— (1972) *On Time and Being* (trans. Joan Stambaugh), New York: Harper & Row.

—— (1982) *Basic Problems of Phenomenology* (trans. Albert Hofstadter), Bloomington, IN: Indiana University Press.

—— (1984) *Early Greek Thinking: The Dawn of Western Philosophy* (trans. David Farrell Krell and Frank A. Capuzzi), San Francisco, CA: Harper & Row.

—— (1985) *History of the Concept of Time: Prolegomena* (trans. Theodore Kisiel), Bloomington, IN: Indiana University Press.

—— (1992) *Parmenides* (trans. André Schuwer and Richard Rojcewicz), Bloomington, IN: Indiana University Press.

Hesiod (1982) *Theogony* (trans. Norman O. Brown), Indianapolis, IN: Bobbs-Merrill.

Homer (1967) *The Iliad* (trans. Richmond Lattimore), Chicago, IL: University of Chicago Press.

Husserl, Edmund (1970) *Cartesian Meditations: An Introduction to Phenomenology* (trans. Dorion Cairns), The Hague: Martinus Nijhoff Publishers.

—— (1981) "Philosophy as rigorous science," in Peter McCormick and Frederick Elliston (eds.) *Husserl: Shorter Works*, Notre Dame, IL: University of Notre Dame Press.

—— (1982) *Ideas Pertaining to a Pure Phenomenology and to a Phenomenological Philosophy* (trans. F. Kersten), The Hague: Martinus Nijhoff Publishers.

—— (1991) *On the Phenomenology of the Consciousness of Internal Time (1893–1917)* (trans. John Barnett Brough), Dordrecht: Kluwer Adademic Publishers.

Hyman, Arthur and Walsh, James J. (eds.) (1983) *Philosophy in the Middle Ages: The Christian, Islamic, and Jewish Traditions*, 2nd edn., Indianapolis, IN: Hackett Publishing Company.

Bibliography

Ishiguro, Hidé (1972) "Leibniz's theory of the ideality of relations," in H. G. Frankfurt (ed.) *Leibniz: A Collection of Critical Essays*, Notre Dame, IL: University of Notre Dame Press.

Jarrett, Charles E., King-Farlow, John and Pelletier, F. J. (1978) *New Essays on Rationalism and Empiricism*, Supplementary vol. IV, Guelph, Ontario: Canadian Association for Publishing in Philosophy.

Jaspers, Karl (1962) *Plato and Augustine* (trans. Ralph Manheim), New York: Harcourt, Brace & World, Inc.

Jordan, Robert (1955) "Time and contingency in St. Augustine," *Review of Metaphysics* 8: 394–417.

Jordan, Robert Welsh (1974) "Intentionality in general," in John Sallis (ed.) *Research in Phenomenology*, vol. 4, Atlantic Highlands, NJ: Humanities Press.

Kant, Immanuel (1956) *Critique of Practical Reason* (trans. Lewis White Beck), Indianapolis, IN: Bobbs-Merrill Company, Inc.

—— (1960) *Religion Within the Limits of Reason Alone* (trans. Theodore M. Greene and Hoyt H. Hudson), New York: Harper & Brothers.

—— (1965) *Critique of Pure Reason* (trans. Norman Kemp Smith), New York: St. Martins Press.

Kirk, G. S., Raven, J. E. and Schofield, M. (1974) *The Presocratic Philosophers: A Critical History with Selected Texts*, Cambridge: Cambridge University Press.

Kline, Morris (1972) *Mathematical Thought: From Ancient to Modern Times*, 3 vols., Oxford: Oxford University Press.

Klossowski, Pierre (1977) "Nietzsche's experience of the eternal return," in David B. Allison (ed.) *The New Nietzsche: Contemporary Styles of Interpretation*, New York: Delta Publishing Co.

Kojève, Alexandre (1969) *Introduction to the Reading of Hegel: Lectures on the Phenomenology of Spirit* (trans. James H. Nichols, Jr.), Ithaca NY: Cornell University Press.

Koyré, Alexandre (1957) *From the Closed World to the Infinite Universe*, New York: Harper Torchbooks.

Kraus, Oskar (1976) "Towards a phenomenognosy of time-consciousness," in Linda L. McAlister (ed.) *The Philosophy of Franz Brentano*, London: Duckworth.

Kripke, Saul (1971) "Identity and necessity," in M. K. Munitz (ed.) *Identity and Individuation*, New York: New York University Press.

Lacey, A. R. (1989) *Bergson*, London: Routledge.

Landes, David S. (1983) *Revolution in Time: Clocks and the Making of the Modern World*, Cambridge, MA: Belknap Press.

Le Goff, Jacques (1988) *Your Money or Your Life: Economy and Religion in the Middle Ages* (trans. Patricia Ranum), New York: Zone Books.

Le Poidevin, Robin (1991) *Change, Cause, and Contradiction: A Defense of the Tenseless Theory of Time*, New York: St. Martin's Press.

Bibliography

Le Poidevin, Robin and MacBeath, Murray (eds.) (1993) *The Philosophy of Time*, Oxford: Oxford University Press.

Le Poidevin, Robin and Mellor, D. H. (1987) "Time, change, and the 'indexical fallacy'," *Mind* 96: 534–538.

Leibniz, Gottfried Wilhelm (1902) *Discourse on Metaphysics, Correspondence with Arnauld, Monadology* (trans. George Montgomery), La Salle, IL: Open Court Publishing Co.

—— (1981) *Leibniz: New Essays on Human Understanding* (ed. and trans. Peter Remnant and Jonathan Bennett), Cambridge: Cambridge University Press.

—— (1989) *Philosophical Essays* (trans. Roger Ariew and Daniel Garber), Indianapolis, IN: Hackett Publishing Co.

Lloyd, G. E. R. (1976) "Views of time in Greek thought," in *Cultures and Time*, Paris; Unesco Press.

Locke, John (1959) *An Essay concerning Human Understanding*, 2 vols., New York: Dover Publications, Inc.

Lorentz, H. A. (1952) "Michaelson's interference experiment," in H. A. Lorentz, A. Einstein, H. Minkowski and H. Weyl, *The Principle of Relativity: A Collection of Original Memoirs on the Special and General Theory of Relativity*, New York: Dover Publications, Inc.

Lorentz, H. A., Einstein, A., Minkowski, H. and Weyl, H. (1952) *The Principle of Relativity: A Collection of Original Memoirs on the Special and General Theory of Relativity*, New York: Dover Publications, Inc.

Lowe, E. J. (1987a) "The indexical fallacy in McTaggart's proof of the unreality of time," *Mind* 96: 62–70.

—— (1987b) "Reply to Le Poidevin and Mellor," *Mind* 96: 539–542.

—— (1992) "McTaggart's paradox revisited," *Mind* 101: 323–326.

Lucretius, Carus Titus (1975) *De Rerum Natura* (Latin with English trans. by W. H. D. Rouse), Loeb Classical Library (Latin authors), no. 181, Cambridge, MA: Harvard University Press.

Mabbot, J. D. "Our direct experience of time," in R. M. Gale (ed.) *The Philosophy of Time: A Collection of Essays*, Garden City, NY: Anchor Doubleday & Company, Inc.

McCall, Storrs (1976) "Objective time flow," *Philosophy of Science* 43: 337–362.

McCormick, Peter and Elliston, Frederick (eds.) (1981) *Husserl: Shorter Works*, Notre Dame, IL: University of Notre Dame Press.

McDowell, John (1978) "On the reality of the past," in Christopher Hookway and Philip Pettit (eds.) *Action and Interpretation: Studies in the Philosophy of the Social Sciences*, Cambridge: Cambridge University Press.

Macintosh, J. J. (1978) "Berkeley's views on time," in Charles E. Jarrett, John King-Farlow and F. J. Pelletier, *New Essays on Rationalism and Empiricism*, Supplementary vol. IV, Guelph, Ontario: Canadian Association for Publishing in Philosophy.

Mackay, D. S. (1935) "Succession and duration," in Stephen C. Pepper *et al.*, *The Problem of Time*, Berkeley, CA: University of California Publications.

McTaggart, John Ellis (1908) "The unreality of time," *Mind*, New Series, no. 68: 457–474.

—— (1927) *The Nature of Existence*, vols. I and II (ed. C. D. Broad), Cambridge: Cambridge University Press.

Maimonides, Moses (1963) *The Guide of the Perplexed* (trans. S. Pines), Chicago, IL: University of Chicago Press.

Marcus, Ruth Barcan (1961) "Modalities and intensional languages," *Synthèse* 130: 303–322.

Marhenke, Paul (1935) "McTaggart's analysis of time," in Stephen C. Pepper *et al.*, *The Problem of Time*, Berkeley, CA: University of California Publications.

Massignon, Louis (1951) "Time in Islamic thought," in *Man and Time: Papers from the Eranos Yearbooks*, vol. 3, New York: Pantheon Books.

Mates, Benson (1986) *The Philosophy of Leibniz: Metaphysics and Language*, New York: Oxford University Press.

Mayo, Bernard (1950) "Events and language," *Analysis* 10(5): 109–114.

Mellor, D. H. (1981) *Real Time*, Cambridge: Cambridge University Press.

Mink, Louis O. (1960) "Time, McTaggart and Pickwickian language," *Philosophical Quarterly* 10: 252–263.

Moore, A. W. (1990) *The Infinite*, London: Routledge.

Moore, F. C. T. (1996) *Bergson: Thinking Backwards*, Cambridge: Cambridge University Press.

Moravcsik, J. M. E. (ed.) (1968) *Aristotle: A Collection of Critical Essays*, London: Macmillan.

Mumford, Lewis (1934) *Technics and Civilization*, New York: Harcourt Brace & Co.

Mundle, C. W. K. (1959) in Paul A. Schilpp (ed.) *The Philosophy of C. D. Broad*, La Salle, IL: Open Court.

Newton, Isaac (1952) *Mathematical Principles of Natural Philosophy*, Chicago, IL: Encyclopaedia Britannica, Inc.

—— (1962) *Unpublished Scientific Papers of Isaac Newton* (ed. A. Rupert Hall and Marie Boas Hall), Cambridge: Cambridge University Press.

—— (1969) *The Mathematical Papers of Isaac Newton*, vol. 3: 1670–1673, New York: Cambridge University Press.

Newton-Smith, W. H. (1980) *The Structure of Time*, London: Routledge & Kegan Paul.

Nietzsche, Friedrich (1966) *Beyond Good and Evil* (trans. Walter Kaufmann), New York: Vintage Books.

—— (1967) *On the Genealogy of Morals and Ecce Homo* (trans. Walter Kaufmann), New York: Viking Press.

—— (1968a) *The Portable Nietzsche* (trans. Walter Kaufmann), New York: Viking Press.

—— (1968b) *The Will To Power* (trans. Walter Kaufmann), New York: Vintage Books.

—— (1974) *The Gay Science* (trans. Walter Kaufmann), New York: Vintage Books.

—— (1986) *Human, All Too Human: A Book for Free Spirits* (trans. R. J. Hollingdale), Cambridge: Cambridge University Press.

Oaklander, L. Nathan (1994) "McTaggart's paradox revisited," in L. Nathan Oaklander and Quentin Smith (eds.) *The New Theory of Time*, New Haven, CT: Yale University Press.

Oaklander, L. Nathan and Smith, Quentin (eds.) (1994) *The New Theory of Time*, New Haven, CT: Yale University Press.

Odegard, D. (1978) "Locke and the specious present," in Charles E. Jarrett, John King-Farlow and F. J. Pelletier, *New Essays on Rationalism and Empiricism*, Supplementary vol. IV, Guelph, Ontario: Canadian Association for Publishing in Philosophy.

Owen, G. E. L. (1976) "Aristotle on time," in Peter Machamer and Robert Turnbull (eds.) *Motion and Time, Space, and Matter: Interrelations in the History and Philosophy of Science*, Ohio: Ohio State University Press.

Pears, D. F. (1956) "Time, truth, and inference," in A. G. N. Flew (ed.) *Essays in Conceptual Analysis*, London: Macmillan & Co.

Pepper, Stephen C. *et al.* (1935) *The Problem of Time*, Berkeley, CA: University of California Publications.

Perry, John (1979) "The problem of the essential indexical," *Nous* 13: 3–21.

Plato (1961) *Plato Collected Dialogues: Including the Letters* (ed. Edith Hamilton and Huntington Cairns), Princeton, NJ: Princeton University Press.

Plotinus, (1991) *The Enneads* (trans. Stephen MacKenna), London: Penguin Books.

Prior, Arthur N. (1958) "Time after time," *Mind* 67: 244–246.

—— (1959) "Thank goodness that's over," *Philosophy* 34: 12–17.

—— (1967) *Past, Present and Future*, Oxford: Oxford University Press.

—— (1968) *Papers on Time and Tense*, Oxford: Oxford University Press.

Quine, Willard Van Orman (1976) "Mr. Strawson on logical theory," in *The Ways of Paradox and Other Essays*, Cambridge, MA: Harvard University Press.

Rankin K. W. (1958) "Order and disorder in time," *Mind* 66: 363–378.

—— (1981) "McTaggart's paradox: two parodies," *Philosophy* 56: 333–348.

Reichenbach, Hans (1966) *Elements of Symbolic Logic*, New York: Free Press.

Rist, J. M. (1969) *Stoic Philosophy*, Cambridge: Cambridge University Press.

Russell, Bertrand (1903) *Principles of Mathematics*, 2nd edn., New York: W. W. Norton & Co., Inc.

—— (1989) *A Critical Exposition of the Philosophy of Leibniz*, Wolfeboro, NH: Longwood Academic.

Samburgsky, S. (1987a) *The Physical World of the Greeks* (trans. Merton Dagut), Princeton, NJ: Princeton University Press.

—— (1987b) *The Physical World of Late Antiquity*, Princeton, NJ: Princeton University Press.

—— (1987c) *Physics of the Stoics*, London: Routledge & Kegan Paul.

Samburgsky, S. and Pines, S. (1971) *The Concept of Time in Late Neoplatonism: Texts with Translation, Introduction and Notes*, Jerusalem: Israel Academy of Sciences and Humanities.

Scheurer, P. B. and Debrock, G. (1988) *Newton's Scientific and Philosophical Legacy*, Dordrecht: Kluwer Academic Publishers.

Schilpp, Paul A. (1959) *The Philosophy of C. D. Broad*, La Salle, IL: Open Court.

Schlesinger, George (1980) *Aspects of Time*, New York: Hackett Publishing Co., Inc.

Schrader, George (1951) "The transcendental ideality and empirical reality of Kant's space and time," *Review of Metaphysics* IV(4): 507–536.

Sextus Empiricus (1990) *Outlines of Pyrrhonism* (trans. R. G. Bury), Buffalo, NY: Prometheus Books.

Shahan, Robert W. and Biro, J. I. (1978) *Spinoza: New Perspectives*, Norman, OK: University of Oklahoma Press.

Sharif, M. M. (ed.) (1963, 1966) *A History of Muslim Philosophy: With Short Accounts of Other Disciplines and the Modern Renaissance in Muslim Lands*, 2 vols., Wiesbaden: Otto Harrassowitz.

Shoemaker, Sidney (1969) "Time without change," *Journal of Philosophy* LXVI: 363–381.

Shorter, J. M. (1984) "The reality of time," *Philosophia* 14: 321–340.

Simplicius (1992) *Corollaries on Place and Time* (trans. J. O. Urmson, annotated by Lucas Siorvanes), Ithaca, NY: Cornell University Press.

Smart, J. J. C. (1956) "The river of time," in A. G. N. Flew (ed.) *Essays in Conceptual Analysis*, London: Macmillan & Co.

—— (1963) *Philosophy and Scientific Realism*, London: Routledge & Kegan Paul.

Smith, Quentin (1994a) "The infinite regress of temporal attributions," in L. Nathan Oaklander and Quentin Smith (eds.) *The New Theory of Time*, New Haven, CT: Yale University Press.

—— (1994b) "The logical structure of the debate about McTaggart's paradox," in L. Nathan Oaklander and Quentin Smith (eds.) *The New Theory of Time*, New Haven, CT: Yale University Press.

Sokolowski, Robert (1964) *The Formation of Husserl's Concept of Constitution*, The Hague: Martinus Nijhoff.

—— (1974) *Husserlian Meditations*, Evanston, IL: Northwestern University Press.

Sorabji, Richard (1983) *Time, Creation and the Continuum: Theories in Antiquity and the Early Middle Ages*, Ithaca, NY: Cornell University Press.

Spinoza, B. (1992) *Ethics, Treatise on the Emendation of the Intellect and Selected Letters* (trans. Samuel Shirley), Indianapolis, IN: Hackett Publishing Co.

—— (1905) *The Principles of Descartes' Philosophy*, La Salle, IL: Open Court Publishing Co.

Stambaugh, Joan (1972) *Nietzsche's Thought of Eternal Return*, Baltimore, MD: Johns Hopkins University Press.

—— (1987) *The Problem of Time in Nietzsche* (trans. John F. Humphrey), London: Bucknell University Press.

Stebbing, L. Susan (1936) "Some ambiguities in discussions concerning time," in R. Klibansky and H. J. Paton (eds.) *Philosophy and History*, Oxford: Clarendon Press.

Stein, Howard (1968) "On Einstein–Minkowski space–time," *Journal of Philosophy* 65: 5–23.

Stobaeus, Ioannes (1860) *Eclogae* (ed. A. Meinecke), Leipzig: Teubner.

Strong, Edward W. (1935) "Time in operational analysis," in Stephen C. Pepper *et al.*, *The Problem of Time*, Berkeley, CA: University of California Publications.

Taylor, Charles (1975) *Hegel*, Cambridge: Cambridge University Press.

Taylor, Richard (1955) "Spatial and temporal analogies and the concept of identity," *Journal of Philosophy* 52: 599–611.

Toulmin, Stephen (1959) "Criticism in the history of science: Newton on absolute space, time, and motion," I and II, *Philosophical Review* 68: 1–29, 203–227.

Wheelwright, Philip (1968) *Heraclitus*, New York: Atheneum.

Whitrow, G. J. (1980) *The Natural Philosophy of Time*, 2nd edn., New York: Oxford University Press.

Williams, Donald C. (1967) "The myth of passage," in R. M. Gale (ed.) *The Philosophy of Time: A Collection of Essays*, Garden City, NY: Anchor Doubleday & Company, Inc.

Wisdom, John (1929) "Time, fact and substance," *Proceedings of the Aristotelian Society* 29: 67–94.

Wolfson, Harry Austryn (1919) "Note on Crescas' definition of time," *Jewish Quarterly Review* X: 1–17.

—— (1958) *The Philosophy of Spinoza*, New York: Meridan Books, Inc.

Wood, David (1989) *The Deconstruction of Time*, Atlantic Highlands, NJ: Humanities Press International, Inc.

Zerubavel, Eviatar (1981) *Hidden Rhythms: Schedules and Calendars in Social Life*, Chicago, IL: University of Chicago Press.

Index

Index

Index

56–9, 62–3, 65–76 *passim*, 81, 83–4, 87, 97–8, 101, 104–5, 111, 114, 142, 145–6, 151–3, 156–7, 160, 169–70, 175, 197

naturalism, naturalistic 8, 57, 158–9

natural science 20, 137–9, 145, 153, 155–8, 200–1

necessity 6–7, 86–8, 91, 95, 114

negation 10, 94, 103, 105, 107–8, 110, 115–16, 224, 227

Neoplatonism, Neoplatonic 43–55, 57–8, 64, 66, 70, 73, 84, 110, 126, 149, 207

new B-theory 145–6, 154

Newton, Isaac, Newtonian 20, 44, 68, 70–85 *passim*, 88–9, 96, 111, 149–50

Nicene formula 57

Nietzsche, Friedrich 107–16 *passim*, 118, 217–18

nihilism 107, 110, 115

no-longer-now 170, 186–7

non-perception 164, 172

nonsense and sense 222–9

nothing 103–4, 107–8, 111, 197

not-yet-now 170, 186–7

noumenal 99, 101–2

novelty 113, 115, 118, 162, 164–5, 168, 170–1, 173, 176–7, 180, 182, 196–8, 202, 205, 207–8, 212, 215, 217–18, 221, 228–9

now, the, nows 22–5, 33–6, 52, 54, 105–6, 118, 129–30, 145–6, 150–3, 163–8, 172–4, 177, 184–7, 192, 215; actually present 162, 166, 177, 181–2; moving 128–30, 152; now point 52–3, 126, 128, 162–4, 168, 170, 173, 177, 181, 185, 187; nows, sequence of 23–4, 33, 105, 185–8, 191; *see also* present

number 15–16, 19–25, 27, 34, 44–5, 50, 64–5, 72, 81, 95, 97, 109, 112, 132, 185, 194–6

numerical distinction 79–80, 83–4, 195, 219

objectivation 174, 180–2

object 10, 75, 86–7, 89–92, 94–7, 99,

103, 105–6, 116, 119, 159–60, 172, 175, 182–3, 186, 193, 200, 202, 222; enduring 166, 178, 181, 198; in general 94–7, 224; immanent 168–9; impossible 226; partial 220–1; physical 12, 175–9, 198; unperceived 204

objectivity 14, 22, 38, 63, 69–71, 78–9, 87, 91, 95–7, 99–100, 102, 106, 119–20, 125, 128, 140, 142–3, 145–6, 153–4, 156–7, 159–60, 166, 169–70, 183, 202, 216–17, 223–5, 22

objective time 22, 63, 69–70, 95–6, 118, 125, 127, 142–3, 145–6, 156–7, 159, 170, 176–8, 229

occurrence 109, 111–12, 114–16, 187–8, 227

Ockham, William of 65–6

One, the 46–51, 54, 57

ontology 9–11, 19, 37–8, 42, 43, 46, 54, 73, 81, 102, 105, 109–15 *passim*, 118–19, 121, 126–7, 137, 145–6, 151, 153, 183, 188, 198, 205, 207–8, 221, 225, 227

ontological difference 119, 183–4, 188, 191

opposites 5, 7–8, 106–8, 110, 112

Oresme, Nicole of 67–8

Ouranos 6

paradox 114, 138, 181, 214–15, 223, 225, 227–8; abstract diagram of 225–6; constitutive 214–15; McTaggart's 124, 127–8, 130, 132–6, 167, 179, 198, 215; skeptical 30–4, 61

paradoxical element 220–2, 224–8

Parmenides of Elea 9–12, 16

partial realities 102, 105–8, 125

particulars 39, 66, 86, 90, 94, 105–6, 113, 119, 134, 143, 151, 158, 212, 214–15, 224

passing away, ceasing to be 7–8, 26, 54, 76–8, 103–6, 111–16, 130–1, 150, 159, 198, 203

past 8, 10, 23, 26, 31–3, 39–42, 48–9, 52, 54, 58–62, 72, 93, 98, 104–6, 109, 111–13, 115, 119,

Index